C Programming
for Scientists
& Engineers

Manufacturing Engineering Modular Series

C Programming for Scientists & Engineers

Robert L Wood

Penton
Press

First published in 2002 by

Penton Press
an imprint of Kogan Page Ltd
120 Pentonville Road
London N1 9JN
www.kogan-page.co.uk

© Robert L Wood, 2002

British Library Cataloguing in Publication Data
A CIP record for this book is available from the British Library

ISBN 1 8571 8030 5

Typeset by Saxon Graphics Ltd, Derby
Printed and bound in Great Britain by Biddles Ltd, Guildford and King's Lynn
www.biddles.co.uk

Contents

Introduction

The aim of this book is to provide a rapid introduction to the C
programming language. C is a procedural language and should not
be confused with C++, which requires a significantly different way
of thinking about problems and their solutions. With the explosion
of texts on C++ and other object-oriented languages in recent
years, along with the perception that C++ is somehow a
progression beyond C, it may seem a little strange to consider a
programming book that is not object oriented. I feel that there are
two good reasons for producing this book. Firstly, object-oriented
analysis, design and programming techniques have evolved to
provide interactive software that is extremely good at supporting
complex tasks performed by its users. However, supporting
computer users in this way is just one aspect of software development. Another significant aspect is the support of numerical
analysis and computer-based modelling in a wide range of engineering and other scientific disciplines, where the priority is to solve
equations as quickly as possible. Examples of this are numerous –
the modelling of stress and temperature distributions in the design
of aircraft and automobiles, the modelling of fluid flow in weather
and climate prediction, the modelling of interactions between
molecules and atoms in the engineering of therapeutic drugs and
new materials.

Using computers to perform the calculations in these and many
other technical applications is a very different problem from
enabling a computer user to do almost anything they want in any
sequence. Both problems are very important, but they need
different tools for their solution. Procedural languages, such as C,
are typically more appropriate than object-oriented languages, such

as C++, for engineering and scientific calculations because the resulting programs can make more efficient use of the relevant hardware resources.

Having said that, the second reason for learning C is that C++ is C with added functionality and that around 90% of any C++ program is actually C. The big difference between C and C++ is not so much in the languages, but in how we think about problems and their solutions. Having thought in an object-oriented way, C++ has the additional functionality over C that allows us to build software that is consistent with our object-oriented thinking. Knowledge of C provides around 90% of the programming knowledge needed to implement object-oriented software.

The approach adopted throughout this book is biased towards generality, rather than comprehensive detail. To this end, this book does not cover every feature that C provides. The decision over what to include and exclude in an introductory text such as this can only be subjective. I apologize to anyone who feels that I have done programming, and C in particular, a disservice by excluding something that they feel strongly about. My main consideration in creating and using these notes has always been to provide a firm foundation on which more specialized knowledge and expertise can be built.

The book is divided into the following chapters:

- variables, data types and declaration statements;
- executable statements;
- functions;
- decisions and loops;
- files and formatting;
- dynamic memory management and linked lists.

Each chapter is further divided into sections that involve the reader in various programming activities guided by tutorial questions. There are further tutorial problems at the end of the book that aim to integrate each chapter topic into the wider framework of C programming. By adopting this approach, it is intended that the reader can learn C through a series of small programming tasks that become incrementally more sophisticated. This incremental development is also used to instill the ideas of writing and using re-usable functions so that, whilst the tutorial questions become more sophisticated, they do not necessarily become more complex or time consuming in their solution.

From this, it should be clear in the reader's mind that this book's main philosophy is that the only way to learn a programming language is to use it. If the reader does not carry out the programming tasks, at best they will only gain a limited understanding of what is possible in C. To understand and use C to write programs that work, it is very important that these tutorial exercises are carried out. In support of these exercises, it is worth noting that this book is independent of any specific programming environment, although all of the tutorial questions have been implemented in both the Borland and Microsoft C/C++ environments.

A further point concerns text style. All examples of C are shown in *italics*, whereas all descriptive text looks like what you are reading now.

For almost a decade, the material in this book has been the basis of both first and second year undergraduate modules, a block-taught (1 week) postgraduate module and a 2-week industrial course. It may seem unusual that a single text should be useful in such a broad range of delivery, but this has been possible due to the way in which the material has been structured. The short, but adequately detailed descriptions of how various C features work, together with frequent opportunities to test new knowledge through practical programming exercises, makes the material attractive to block and short course teaching beyond undergraduate level. Under these regimes, all parts of these notes have been mandatory and assessment has involved the design and programming of software to solve significant technical problems, such as the thermodynamic modelling of a whole engine cycle. In contrast, at the undergraduate level, knowledge of Chapter 6 concerning dynamic memory management is not expected and there is less emphasis on the integrating nature of the tutorial questions at the end of the book. Also, assessment problems are relatively small, but still of a technically applied nature.

Now it is time to get a little more focused on the subject at hand. The following comments are intended to introduce a few important C words and make clear the relationships between them.

All C programs contain statements. The programmer assembles these by combining various operators, such as '*add*', '*divide*' etc., and variables, such as X or Y. There are two general types of statements in C – declaration statements that are used to create variables, and executable statements used to combine operators and variables in ways that make the computer do something useful. In all but the smallest of programs, the programmer should package or group statements related to a particular task into functions. For example, a program that must read a collection of input data, perform calculations and output the results could contain a function for each of these tasks. Every C program has a function called *main*, which is always the first part of the program to run. Very small programs, including many of the examples in this book, may contain so few statements that they can all reasonably be contained in *main*. In larger programs, *main* typically calls or uses other functions to carry out particular tasks. The C language provides many standard functions that perform specific tasks, such as reading a value from the keyboard, calculating a square root, etc. These standard functions are grouped into libraries and, to use them, it is necessary to have a *#include* statement that refers to the relevant library at the start of the program.

Where C does not provide a suitable function for the programmer's need, the programmer must create one. This is what C programming is about – understanding how to combine C operators with variables to form statements, and to group the statements into appropriate functions. Looking back at the list of chapters, the first three chapters are intended to support this by concentrating on the terms mentioned above. All of the programs in these chapters are limited to reading data from the keyboard, sometimes carrying out a simple sequence of instructions, and displaying results on the screen. Subsequent chapters build on this basic functionality by introducing some of the more sophisticated facilities that C provides. For example, Chapter 4 takes an important step forward by looking at how C programs can make decisions, such as '*if* ...', or repeat sets of statements within *while* and *for* loops. Another step forward is to look at how a C program can work with files, in addition to the keyboard and screen. The final step taken in this book is concerned with how C programs can create their own variables in the form of linked lists.

One final comment for readers who have never programmed before. In C (and all other programming languages) there are quite a lot of rules that dictate how statements and functions can be constructed. For example, all C statements must end with a semicolon ';'. Also, when a variable is created it must be given a name. Wherever a variable is used in a program, its name must be spelled in exactly the same way as it is in its declaration statement. C allows both upper and lower case symbols to be used in the names of variables but, again, their use must be the same wherever the variable is used. A good knowledge of these (and a few more) rules is an important element of successful programming and, perhaps, the main difference between just reading a programming book and working through the exercises that it contains. In the early stages of learning any programming language, you need to recognize that you will make some mistakes and have to deal with error and warning messages issued by your programming environment. The light at the end of the tunnel, however, is that the more attention you pay to detail, the quicker the error messages will go away.

1

Variables, Data Types and Declaration Statements

1.1 Introduction

All programs work with data stored in memory. To make it easier for us to design and understand programs, we give each data item a unique name. By doing this we do not need to know where the data item is stored in memory, we just use its name wherever it is needed in the program. This is very similar to what we do when we write a mathematical equation, for example $x = y + z$, where x, y and z are the names of three variables. In a C program, x, y and z would also be the names of three variables, each stored at a different location or **address** in memory. The program needs to know where but, generally, we as programmers do not. Thus, a **variable** is an item of data that has a name and whose value can be changed within a program. Before a **variable** can be used, it must be given a name and declared to be of some specific **data type**. This is done in a **declaration statement**. In C there are four basic data types:

- Character, e.g. 'a', 'b', 'c', etc.
- Integer, e.g. 1, 2, 3, etc.
- Real, e.g. 1.0, 2.0, 3.0, etc.
- Pointer.

Whilst the meaning of the first three types of data is hopefully clear from these examples, the Pointer data type is rather unusual and will be considered later.

The amount of space (the number of bytes) needed in memory to hold a variable depends on its data type. Thus, two bytes may be needed to store an integer type variable and four bytes may be needed to store a variable of type real. In addition to variables of the basic data types shown above, the programmer can also define any group, set or aggregate of these variables. An **array** is used to hold a collection of variables where all of the variables are of the same data type. The programmer can also create **data structures**, built up from various combinations of the basic data types, arrays and other data structures. Data structures have another special significance in C because C treats them as programmer-defined data types.

Sections in this chapter consider variables of each data type, above, showing how they are created using declaration statements and how they are used to store data that is read from the keyboard and then displayed on the screen.

1.2 The character data type

C stores characters in memory as integer numbers using the ASCII code.[1] Every number in the ASCII code is small enough to be stored in a single byte. Hence, a variable of type character uses one byte. Variables of the character data type are declared using statements such as:

char A;   declares a variable called *A* to hold one character

char symbol, letter;   declares variables *symbol* and *letter* to each hold
                               one character

In a declaration statement *char* defines the character data type and is followed by the name(s) of the required variable(s) separated by commas. Remember from the Introduction that C requires a semi-colon, ;, at the end of each statement. Variables of type *char* can be given a value, or initialized in a declaration statement using single quotes, as follows:

char A = 'a', B = 'd';
char C = 'M';

Variables of type *char* can only hold a single character. To hold a **character string**, such as a person's name, an array of type *char* is

---

[1]ASCII = American Standard Code for Information Interchange.

required. Arrays and character strings are introduced in Sections 1.6 and 1.7, respectively.

Character data can be read from the keyboard using the *fscanf* function and written to the screen using the *fprintf* function. For example, Program 1.1 reads a single character from the keyboard and displays it on the screen.

```
/* Program 1.1  -  Reading and writing a character */

#include <stdio.h>

int main(void)
{
char A;
fprintf(stdout, "Enter a single character:");
fscanf(stdin, "%c", &A);
fprintf(stdout, "The character is %c\n", A);
return(0);
}
```

The listing for Program 1.1 starts with a comment, giving some indication of what the program does. The *#include <stdio.h>* statement will appear in the first group of statements in every complete program in these notes. This statement is needed in a program if it uses the *fprintf* or *fscanf* functions. More of the detail behind this will be revealed in Chapter 3. Inside the program *char A;* declares a variable, *A*, which can hold a single character. To make sure that the user knows that they must type a character, the program sends a message, *'Enter a single character:'*, to the screen using the *fprintf* function. The first **argument** (item of information), *stdout*, supplied to *fprintf* is a **stream** that connects the program to the screen. When we use this stream, we are telling *fprintf* to send the message to the screen, rather than to some other part of the computer, such as a file on disc. Streams are discussed further in Chapter 5. For now, however, simply remember that the *stdout* stream always connects a program to the screen.

When the above program has displayed the message on the screen, it then calls the *fscanf* function to read data from the keyboard. The *fscanf* function uses three arguments. The first is a stream, *stdin*, which always connects a program to the keyboard. The second argument is the **control string**, *'%c'*. The *%c* part of the control string in this example is a **formatting code** that instructs *fscanf* to interpret the data that it reads from the keyboard as a character. The third argument, *&A*, instructs *fscanf* to store the character that it has read in a variable called *A*. It is very important to note that the *&* symbol has been used in front of the name of the variable. The *&* symbol is called the '**address of**' operator. When it is put in front of a variable it gets the location in memory where the variable is stored. Hence, *&A* should be read as 'the address (in memory) of the variable called *A*'. The *&* operator is needed when we use *fscanf* because *fscanf* can only put the data that it reads into specific places in memory. More importantly, *fscanf* needs to be told where to put the data that it reads. The easiest way to do this is to declare a variable beforehand and to say 'put the data at the address of that variable'. The overall effect of this is to store the character supplied from the keyboard in the variable *A*.

When *fscanf* has done its job, the program then calls the *fprintf* function again. This time, *fprintf* has to display a message on the screen that contains the contents or value of variable *A*. The message to be displayed is inside the control string, which is the second argument. Inside the control string, the formatting code, *%c*, indicates where the value of *A* will be inserted into the message and that the value to be inserted is of type *char*. The control string also contains the symbols '*\n*', which together are a **control code** that forces the cursor to go to the start of the next line on the screen. The third argument specifies the variable, *A*, whose value is to replace *%c* in the message.

Tutorial 1.1
Implement Program 1.1. Write brief notes on the action of each statement in the program.

Tutorial 1.2
Modify Program 1.1 to store the character data in a variable called *character_data*.

1.3 The integer data type

Integer type variables are used to store whole numbers and are declared using statements such as:

> int A, B;   declares two variables called A and B to each hold one integer
> value

In a declaration statement *int* specifies the integer data type and is followed by the name(s) of the required variable(s), separated by commas. By default, C allocates a fixed number of bytes in which to store an integer value. This places a default upper limit on the magnitude of the values that can be stored. If a value greater than this default is required, *int* can be preceded by *long* in a declaration statement. This tells C to use more bytes for the variable. Conversely, if the maximum value to be stored is smaller than the default maximum, it may be possible to save memory by using *short int*. Another feature of the integer data type is that variables may hold either positive (unsigned) values only or either positive or negative values (signed). To restrict an integer variable to storing only positive values, *int* is preceded by *unsigned*. The ANSI[2] standard data types for these different options are shown in Table 1.1, along with the amount of memory used and the minimum range of values that can be stored:

Table 1.1 *Integer data types*

Data type	Memory (bytes)	Value range
short int	2	–32,768 to 32,767
unsigned short int	2	0 to 65,535
int	2	–32,768 to 32,767
unsigned int	2	0 to 65,535
long int	4	–2,147,483,648 to 2,147,483,647
unsigned long int	4	0 to 4,294,967,295

In contrast to the above minimum values, some compilers and processors allocate greater amounts of memory to the *int* and *long int* data types, allowing a correspondingly greater range of values to

---

[2]ANSI = American National Standards Institute.

be stored. In addition to storing integer values in *int* type variables, *char* variables can also be used. However, since *char* variables occupy just one byte, the following restrictions apply, depending on whether the variable is signed or unsigned:

char	integer character code, range 0 to 127
signed char	signed integer values within the range –128 to 127
unsigned char	integer values within the range 0 to 255

Typical forms of declaration statements for integers are:

int A;	declares an *int* variable called *A*
int counter, limit = 100;	declares two *int* variables, initializing the second
short int B = –32000;	declares and initializes a *short int* variable

Since there are several types of integer variables, different formatting codes are required by *fscanf* to read data into them from the keyboard, and by *fprintf* to display their value on the screen. The formatting codes that are required for integer data often consist of two symbols after the % sign. When two symbols are required, the first is a **modifier** and the second is the basic formatting code. There are two basic formatting codes: *d* for integer values that may be positive or negative and *u* for values that are positive (unsigned) only. In addition, there are two modifiers, *h* meaning *short* and *l* meaning *long*. Table 1.2 summarizes these options.

Table 1.2 *Formatting codes required for different integer data types*

Data type	Basic formatting code	Modifier	Required formatting code
short int	%d	*h*	%hd
unsigned short int	%u	*h*	%hu
int	%d		%d
unsigned int	%u		%u
long int	%d	*l*	%ld
unsigned long int	%u	*l*	%lu

Program 1.2 shows how different formatting codes are used when the various integer data types are read from the keyboard using *fscanf* and displayed on the screen using *fprintf*.

/* Program 1.2 - Reading and writing different types of integer data */

```
#include <stdio.h>

int main(void)
{
short int A;
unsigned short int B;
int C;
unsigned int D;
long int E;
unsigned long int F;

/* reading and writing a short int */
fprintf(stdout, "Enter an integer value between -32768 and 32767:");
fscanf(stdin, "%hd", &A);
fprintf(stdout, "The integer value is %hd\n", A);

/* reading and writing an unsigned short int */
fprintf(stdout, "Enter an integer value between 0 and 65535:");
fscanf(stdin, "%hu", &B);
fprintf(stdout, "The integer value is %hu\n", B);

/* reading and writing an int */
fprintf(stdout, "Enter an integer value between -32768 and 32767:");
fscanf(stdin, "%d", &C);
fprintf(stdout, "The integer value is %d\n", C);

/* reading and writing an unsigned int */
fprintf(stdout, "Enter an integer value between 0 and 65535:");
fscanf(stdin, "%u", &D);
fprintf(stdout, "The integer value is %u\n", D);

/* reading and writing a long int */
fprintf(stdout, "Enter an integer value between -2147483648 and 2147483647:");
fscanf(stdin, "%ld", &E);
fprintf(stdout, "The integer value is %ld\n", E);

/* reading and writing an unsigned long int */
fprintf(stdout, "Enter an integer value between 0 and 4294967295:");
fscanf(stdin, "%lu", &F);
```

```
fprintf(stdout, " The integer value is %lu\n", F);

return(0);
}
```

Tutorial 1.3
Implement Program 1.2. Write brief notes on the action of each statement in the program.

Tutorial 1.4
Modify Program 1.2 so that the values that are read in are displayed in reverse order. Ensure that the program contains appropriate comments and that appropriate messages appear on the screen with each displayed value.

1.4 The real data type

Real numbers are often written in decimal form, e.g. 102.7, or in exponential form, e.g. 1.027×10^2. In C the decimal notation is the same, but the exponential notation looks like 1.027e2. As shown in Table 1.3, values of this type can be held in any of three data types in the real category, depending on the required precision (decimal places, d.p.) and the range (the maximum and minimum values) of the variable.

Table 1.3 *Data types in the real category*

Data type	Memory bytes	Range of values	Precision d.p.
float	4	1.175494351e-38 to 3.402823466e+38	7
double	8	2.22507385850720e-308 to 1.79769313486231e+308	15
long double	10	3.36210314311209e-4932 to 1.18973149535723e+4932	19

The *float* and *double* data types are available in all C programming environments, but *long double* is limited to the use of very high precision floating point hardware. Also, variables of type *float* and *double* are often referred to as **single precision** and **double precision** variables, respectively. Typical forms of declaration statement are:

float A;	declares a single precision variable, called *A*
double A = 19.2, B = 1.7e-6;	declares and initializes two double precision variables

Data of types *float* and *double* can be transferred into and out of programs either in decimal or exponential formats, depending on the formatting codes that are used with *fscanf* and *fprintf*. For the *float* data type the formatting code required for decimal format is %*f* and for the *double* data type the %*lf* formatting code is used, where *l* is a modifier. The formatting code %*e* is used for *float* type variables when the data is to be displayed in exponential format. To display the value of a *double* type variable in exponential format %*le* is used. Program 1.3 demonstrates these formatting codes in the input and output of *float* and *double* type variables using *fscanf* and *fprintf*.

```
/* Program 1.3 - Reading and writing floats and doubles */

#include <stdio.h>

int main(void)
{
float A;
double B;

/* reading and writing a float */
fprintf(stdout, "Enter a value between 1.175e-38 and 3.402e+38 as a decimal:");
fscanf(stdin, "%f", &A);
fprintf(stdout, "The value as a decimal is %f\n", A);
fprintf(stdout, "The value as an exponential is %e\n", A);

fprintf(stdout, "Enter a value between 1.175e-38 and 3.402e+38 as an "
"exponential:");
fscanf(stdin, "%e", &A);
fprintf(stdout, "The value as a decimal is %f\n", A);
fprintf(stdout, "The value as an exponential is %e\n", A);
```

```
/* reading and writing a double */
fprintf(stdout, "Enter a value between 2.225e-308 and 1.797e+308 as a decimal:");
fscanf(stdin, "%lf", &B);
fprintf(stdout, "The value as a decimal is %lf\n", B);
fprintf(stdout, "The value as an exponential is %e\n", B);

/* reading and writing a double */
fprintf(stdout, "Enter a value between 2.225e-308 and 1.797e+308 as an"
"exponential:");
fscanf(stdin, "%le", &B);
fprintf(stdout, "The value as a decimal is %lf\n", B);
fprintf(stdout, "The value as an exponential is %le\n", B);

return(0);
}
```

Tutorial 1.5
Implement Program 1.3. Write brief notes on the action of each statement in the program.

Tutorial 1.6
In Program 1.3, the rules for entering data in either decimal or exponential format are not strict. Enter data in different formats and note the resulting output.

1.5 The pointer data type

To understand what pointer variables are it is necessary to reconsider the difference between the value of any variable and the location or address of that variable in memory, outlined in Section 1.1. Remember that, by giving a name to a variable, we are allocating one or more bytes of memory in which we can store an item of data. Also, by working with named variables, we don't need to worry about exactly where the variable (meaning item of data) is located in memory. There are, however, several situations in which we need to work with variables via their locations, rather than using their names. For example, when *fscanf* reads an item of data, it needs to be told where to put that item in memory. In all of the programs that you have seen in this chapter, this has been done by putting '&' in front of the variable name. As mentioned previously, & is called the 'address of' operator. This operator should be thought of as a tool that finds the location of the variable whose name follows it. For example, &*A* finds where *A* is located in memory. In other words, & gets the address of *A*. When used with *fscanf*, you need to imagine that the '&*A*' symbols are replaced by the address of *A* which is then given to *fscanf*. Suppose, instead of passing the address to *fscanf*, we wanted to store it in another variable using a statement such as *Z* = &*A*;. To do this, *Z* would have to be a variable of type pointer. So, the pointer data type allows us to create variables that are used to store the address of other variables. We generally say that pointer variables (or simply 'pointers') hold the address of, or 'point to', other variables.

To understand how programs in later chapters carry out their tasks, it is useful to have a mental picture of how the 'address of' operator gets the address of a variable and how this address can be stored in a pointer variable. Suppose, when an integer variable is declared (e.g. in a statement such as *int A;*) that two consecutive bytes are reserved in memory. This group of bytes is then given the name of the variable, *A*. Information about this variable is stored in a table that the program creates, called a look-up table. Essentially, for any declared variable, the information stored in a look-up table consists of the name of the variable, its data type and the location in memory of the first byte that it uses. The location of the first byte of a variable is called its address. To give another example, a declaration statement such as *double B = 1.75e10;* reserves 8 consecutive bytes in memory and then stores the value *1.75e10* in these bytes. Again, the look-up table will store the name of the variable, *B*, its data type, *double*, and the location in memory of the first byte that has been used (the address of *B*). It is very important to remember that the contents of a variable (its value) and the location where it is stored in memory (its address) are quite different. When the 'address of' operator, &, is used, for example &*B*, the operator is given the name *B* which it searches for in the look-up table. Having found *B*, & then uses the look-up table to retrieve the location in memory of the first byte used to store *B*. It is this location that would be stored in a pointer using a statement such as *Z* = &*B*;, where *Z* needs to have been previously declared as a pointer variable.

Pointer variables must be declared using the same data type as the variables they will hold the addresses of. In other words, the address of an *int* type variable can only be stored in a pointer of type *int*. Similarly, the address of a *float* type variable can only be stored in a pointer of type *float*, etc. The declaration statement for a pointer is very similar to that for other variables, except that an asterisk, *, is used in front of the variable name. For example:

```
int *A;    declares a variable, A, to be a pointer to variables of type int
float *B;  declares a variable, B, to be a pointer to variables of type float
char *C;   declares a variable, C, to be a pointer to variables of type char
double *Z; declares a variable, Z, to be a pointer to variables of type double
```

Having declared a pointer variable, the address of another variable can be stored in it by using the 'address of' operator, &, as in the following example:

```
double B;  declares a variable of type double, called B
double *Z; declares a pointer of type double, called Z
Z = &B;    stores the address of B in pointer Z
```

When a statement such as $Z = \&B;$ is executed, what actually happens is that the address in memory of the first byte of the variable, *B*, is returned by & and stored in *Z*.

The * operator is very important because it can have three meanings, depending on whether it is used in a declaration statement or an executable statement. In a declaration statement, * means that the variable named after it is a pointer. In an executable statement, * can be the multiply operator or the '**contents of**' operator. To use * as the 'contents of' operator, it is put in front of a pointer. This gives the value of the variable whose address is stored in the pointer. For example:

```
int A = 4, C; declares two integer variables, initializing the first
int *B;       declares B to be a pointer of type int
B = &A;       stores the address of A in pointer B
C = *B;       the value of A, pointed to by B, is copied to C (C is given the
              value 4)
```

In this example, the first two statements are declaration statements and the last two are executable statements. Using * in the second declaration statement says that *B* is a pointer of type *int*. Using * as the 'contents of' operator in the second executable statement enables the value of *A* to be retrieved because *B* holds the address of *A*.

Looking back over the example programs in this chapter, you should again note that when *fscanf* is used to read values from the keyboard, the values are stored at the addresses of the variables in the *fscanf* argument list. This has been done by prefixing the name of each variable with the 'address of' operator, *&*. As shown in Program 1.4, this can also be achieved using pointers.

```
/* Program 1.4 - The use of pointer variables in reading and writing */

#include <stdio.h>

int main(void)
{
float A;
float *A_ptr;
int B;
int *B_ptr;

A_ptr = &A;
B_ptr = &B;
fprintf(stdout, "Enter a decimal value:");
fscanf(stdin, "%f", A_ptr);
fprintf(stdout, "The value entered is %f\n", A);
fprintf(stdout, "The value entered is %f\n", *A_ptr);

fprintf(stdout, "Enter an integer value:");
fscanf(stdin, "%d", B_ptr);
fprintf(stdout, "The value entered is %d\n", B);
fprintf(stdout, "The value entered is %d\n", *B_ptr);
return(0);
}
```

Program 1.4 reads a real number and an integer from the keyboard and displays them on the screen. The real value is stored in a *float* type variable, *A*, and the integer value is stored in an *int* type variable *B*. Having declared *A* and *B*, two pointers are also declared, called *A_ptr* and *B_ptr*. *A_ptr* is of type *float* and is used to store the address of variable *A*. Similarly, *B_ptr* is of type *int* and used to store the address of *B*. When *fscanf* is called to read the data from the keyboard, it needs to be given the address of the variables that will be used to store each item of data. In this example, the addresses of *A* and *B* are already stored in *A_ptr* and *B_ptr*, respectively. This means that the addresses needed by *fscanf* can be specified using *A_ptr* and *B_ptr*, rather than *&A* and *&B*. Following each call to *fscanf* there are two calls to *fprintf*, each of which displays the same message on the screen. However, the two calls do this in different ways. The first involves specifying the name of the variable to be displayed. In the second call, the 'contents of' operator prefixes each pointer variable. This means that the contents of the variable pointed to by the pointer will be displayed. More specifically, **A_ptr* will give the value of *A* because *A_ptr* holds the address of *A*. Likewise, **B_ptr* will give the value of *B* because *B_ptr* holds the address of *B*.

Tutorial 1.7
Implement Program 1.4. Write brief notes on the action of each statement in the program.

Tutorial 1.8
Using Program 1.4 as a guide, implement a program that reads values from the keyboard into variables of type *short int*, *long unsigned int* and *double* and then displays them on the screen. Use pointer variables of the correct type to specify where *fscanf* should store the data and use the 'contents of' operator with *fprintf* to display the data.

1.6 Arrays

The preceding paragraphs have introduced the basic data types, typical declaration statements for variables of these types and a general approach to getting individual items of information into and out of programs. However, it is more often the case that a program is required to work with collections of data, ranging from a 'handful' of numbers to thousands or millions of values. To help do this efficiently, C provides facilities to group items of data together and to treat these groups as entities or variables in their own right. This section introduces two alternative and complementary methods for storing such composite data items. One of these methods is based on the use of **arrays** for storing groups of data, where each item in the group is of the same data type. Character strings, such as the name of a person, are a particular type of array and are considered in the next section. The second method of grouping data items, considered in Section 1.8, uses data structures, in which variables of different data types can be grouped together. The C language is very flexible in that it also allows the programmer to create arrays of arrays, arrays of data structures and data structures that contain arrays. Typical examples of array declaration statements are:

int A[10];   an array of 10 integers
double B[5]; an array of 5 real numbers of type *double*
float C[20];   an array of 20 real numbers of type *float*

Each of the above arrays are one-dimensional; hence they are sometimes also called **vectors**. Within an array an individual item of data is called an **element**. Hence a declaration statement such as *double B[5];* declares an array having the five elements *B[0]*, *B[1]*, *B[2]*, *B[3]* and *B[4]*, where each element is a variable of type *double*. Since the elements of an array are stored sequentially in memory, declaration statements such as *double B[5];* use blocks of memory that are divided into array elements. The number of bytes used is given by the number of elements in the array multiplied by the number of bytes for the relevant data type. The previous list of elements in *B* highlights a very important point concerning arrays, in that the numbering of array elements always starts at zero and goes up to the number used in the declaration statement minus one. This numbering convention is ALWAYS used for arrays in C.

In addition to one-dimensional arrays, C also allows the use of arrays having two or more dimensions, for example *int A[3][3];* declares a two-dimensional array having the nine elements:

A[0][0]   A[0][1]   A[0][2]
A[1][0]   A[1][1]   A[1][2]
A[2][0]   A[2][1]   A[2][2]

If needed, arrays can be initialized when they are declared. For example, to store the integer values 1 to 9 in *A* at the time that it is created, we would use:

int A[3][3] = {{1, 2, 3}, {4, 5, 6}, {7, 8, 9}};

resulting in:

```
A[0][0] = 1     A[0][1] = 2     A[0][2] = 3
A[1][0] = 4     A[1][1] = 5     A[1][2] = 6
A[2][0] = 7     A[2][1] = 8     A[2][2] = 9
```

Program 1.5 demonstrates the reading and writing of information using a one-dimensional array.

```
/* Program 1.5 - Reading and writing an array of numbers */

#include <stdio.h>

int main(void)
{
float A[3];

fprintf(stdout, "Enter three numbers:");
fscanf(stdin, "%f %f %f", &A[0], &A[1], &A[2]);
fprintf(stdout, "The numbers entered are\n");
fprintf(stdout, "A[0] = %f  A[1] = %f  A[2] = %f \n", A[0], A[1], A[2]);

return(0);
}
```

In Program 1.5 an array, *A*, is declared to hold three values of type *float*. After *fprintf* is used to prompt the user for input, *fscanf* reads three values from the keyboard. Note that a formatting code is required for each value that is to be read and that the address of each required array element must also be separately specified. After the numbers have been stored in the array, two calls to *fprintf* display them on the screen. Two points to note about the final call to *fprintf* are, firstly, that a formatting code is required for each value to be displayed and, secondly, that the array elements are specified as the 3rd, 4th and 5th arguments, in the correct order for their insertion into the displayed message.

The most useful feature of arrays is that they allow potentially large quantities of data, of the same data type, to be stored, accessed and processed efficiently. However, arrays have three limitations. Firstly, a single array can only hold data values of the same type. In many situations we would like to have the convenience of arrays, but for mixed data types. Secondly, arrays must be specified to have a fixed size (number of elements and dimensions) when they are declared. This often leads to inefficient use of memory and arbitrary limitations on the volume of data that can be stored or processed. The third problem is that arrays occupy contiguous blocks of memory. For example, the array in Program 1.5 occupies only 12 consecutive bytes of memory. However, if the array were declared as *float A[3000];* then it would occupy 12,000 consecutive bytes. Depending on the compiler and operating system that are used, there is always an upper limit on the sizes of individual blocks of memory that can be used. These limits may be smaller than the amount of data that needs to be processed. Although these limitations do not usually present a significant problem in relatively 'small scale' software, they can be very important in large engineering and scientific programs. In such cases, alternative methods of grouping data are to be preferred. The central concepts of these methods will be investigated in Chapter 6.

Tutorial 1.9
Implement Program 1.5. Write brief notes on the action of each statement in the program.

Tutorial 1.10
Modify Program 1.5 so that it reads and writes two arrays. The first array contains three values of type *unsigned int* and the second array contains five values of type *double*.

1.7 Character strings

Character strings are a special type of array. For example:

char A[10];   declares a character string, called *A*, which can hold a **maximum of nine characters**.

The reason why only nine characters can be stored in A is that character strings can consist of any number of bytes, one per character, and it is necessary to indicate where the string finishes in memory. For example, if A, above, is used to store the string *'abc'* the program needs to know that only the first three bytes of A contain useful data. This is done by storing a 'null' character, \0, after the last useful byte in the string. C automatically inserts the null character in the first byte that does not contain a useful character. However, the programmer must declare the string to be large enough to hold the maximum required data AND the null character. Thus, when a character string is declared, the number in the brackets must be at least one bigger than the maximum number of characters to be stored in the string. Typical forms of the declaration statement for character strings are:

char A[11];	declares a string, A, to hold 10 characters
char A[12] = "Hello World";	declares a string, A, to hold 11 characters, initializing it
char A[4]= "abc", B[81];	declares two character strings, initializing the first

In addition to these single character strings, arrays of character strings can also be declared. For example:

char[2][11];	declares a vector of two character strings, each holding up to 10 characters
char words[5][5][21];	declares a two-dimensional array of 25 character strings, where each string can hold up to 20 characters

Program 1.6 demonstrates the reading and writing of a character string.

/* Program 1.6 - Reading and writing a character string */

```
#include <stdio.h>

int main(void)
{
char A[101];

fprintf(stdout, "Enter your name:");
fscanf(stdin, "%s", A);
fprintf(stdout, "Your name is %s\n", A);

return(0);
}
```

In Program 1.6 a character string that can store up to 100 characters is declared. The first executable statement is a call to *fprintf*, which prompts the user to supply their name. The *fscanf* function is then used to read the name from the keyboard. Note that the control string for *fscanf* contains a new formatting code, *%s*. This tells *fscanf* to interpret the input as a character string. This should be compared with the formatting code *%c*, introduced in Section 1.2, required for single characters. A further point to note in the call to *fscanf* in Program 1.6, is that the name of the character string, *A*, is not preceded by *&*. This only works for character strings and is allowed because the name of an array is actually a pointer to the first byte in the array. As an alternative to this we could use the approach demonstrated in Program 1.5, where the address of each array element is specified explicitly. However, this is inefficient for character strings. REMEMBER that C treats character strings differently to variables of other data types in that *fscanf* does not need *&* in front of the name of the character string. Hence, as shown in Program 1.6, both *fscanf* and *fprintf* just need the name of the character string as an argument.

A further point to note when reading data into character strings is that a character string is defined as a continuous sequence of symbols. Thus, 'John' is a single character string, but 'John Smith' consists of two character strings. The result of typing either of these as input to Program 1.6 would be the same, the output on the screen would be 'Your name is John'.

Tutorial 1.11
Implement Program 1.6. Write brief notes on the action of each statement in the program.

Tutorial 1.12
Using Program 1.6 as a guide, implement a program that reads and writes an array of five words, each containing up to 10 characters.

1.8 Data structures

Suppose that we need a program that reads collections of data from the user and displays their values on the screen. For example, each collection could refer to an employee in a company, consisting of their employee number, family name and salary. The most appropriate data types for storing these items are *int* for employee number, *char* for family name and *float* for salary. If this information is to be processed for many employees, we could declare three arrays, one for employee numbers, another for family names and a third for salaries. By doing this, however, the original collection of data for each employee has been split up and stored in different parts of the program. This can give rise to various difficulties, depending on what we intend to do with the data. It is generally better to keep the items of data within a group together and to do this, we need to use data structures. The following is an example of a data structure definition that suits the requirements for employee's data.

```
struct employee
  {
  int number;
  char family_name[101];
  float salary;
  };
```

Every data structure must have a name, for example *struct employee*, above. Whenever a structure is defined, it is treated by the compiler as a new data type, in the same way as *int, float, char*, etc. Hence, the name of the above structure is also the name of a new data type. It is important to be very clear about this. The above statements do not declare or create a variable. They define a new data type that can be used in other declaration statements, just like *int, float, char*, etc., are defined by C to be used in declaration statements. For example, *int P;* declares a variable, *P*, to be type *int*, and *struct employee K;* declares variable, *K*, to be of type *struct employee*. The difference between *P* and *K* is that *P* is a single item of data stored in two bytes, whereas *K* is a collection of three variables that occupies 107 bytes. Program 1.7 uses the employee example to show how data structures are defined and used within a program.

```
/* Program 1.7 - Reading and writing a data structure */

#include <stdio.h>

int main(void)
```

```
{
struct employee
  {
  int number;
  char family_name[101];
  float salary;
  };

struct employee employee_1;

fprintf(stdout, "Enter employee number:");
fscanf(stdin, "%d", &employee_1.number);
fprintf(stdout, "Enter employee family name:");
fscanf(stdin, "%s", &employee_1.family_name);
fprintf(stdout, "Enter employee salary:");
fscanf(stdin, "%f", &employee_1.salary);

fprintf(stdout,"Employee: name is %s   number is %d  salary is %f\n",
                employee_1.family_name,
                employee_1.number,
                employee_1.salary);
return(0);
}
```

In Program 1.7, the first declaration statement defines the *struct employee* data type. The second declaration statement creates a variable, called *employee_1*, of type *struct employee*. This variable is an actual data structure, located somewhere in memory. The memory used is partitioned into three **member variables**, as specified in the definition of *struct employee*. The six executable statements that follow these declarations simply prompt the user to supply each item of data to be stored in *employee_1* and store the data that is supplied. Note that each member of *employee_1* is accessed by specifying the structure name and the member name, separated by a full stop or period. This is the **'dot' operator** which allows access to each member of a structure by **fully qualifying** it. Note also that *fscanf* needs to know the address of each member and this is done by prefixing the fully qualified member names with the *&* operator (including the name of the character string because it is a member of a structure). Having read the necessary data, *main* then calls *fprintf*, to display the data values, again accessing each member of the data structure by fully qualifying it.

If the above data has to be stored for a number of employees, we could use an array of data structures. For example, the statement:

```
struct employee employees[10];
```

creates an array containing 10 elements, each of which is a variable of type *struct employee*. In other words, this statement creates 10 data structures and groups them together in a single array, called *employees*. To access the data for any individual employee, it is necessary to combine the ways of accessing elements of arrays and members of data structures. Thus, the data for the first employee in the array is accessed using the following statements:

```
employees[0].number
employees[0].family_name
employees[0].salary
```

and data for the third employee is accessed using:

```
employees[2].number
employees[2].family_name
employees[2].salary
```

It should be noted, however, that arrays of data structures are subject to the limitations discussed in Section 1.6 concerning their fixed size. Alternative methods of storing multiple groups of data that overcome these limitations will be introduced in Chapter 6.

Tutorial 1.13
Implement Program 1.7. Write brief notes on the action of each statement in the program.

Tutorial 1.14
Using an array of structures, modify Program 1.7 so that it will read, store and display data for three employees, reading all of the data before displaying it.

1.9 Pointers to data structures

In addition to using the dot operator to access members of a data structure by fully qualifying them, **indirect access** is also possible through the use of a pointer and the **'indirection' operator**, '->'. To do this it is necessary to declare a pointer of the correct data type, as shown below.

```
struct employee
  {
  int number;
  char family_name[101];
  float salary;
  };

struct employee employee_1, *employee_1_ptr;
employee_1_ptr = &employee_1;


    employee_1_ptr->number
    employee_1_ptr->family_name
    employee_1_ptr->salary
```

In the above example, having defined *struct employee* as a new data type, *employee_1* is declared to be a variable of type *struct employee*. The same declaration statement also declares a pointer, *employee_1_ptr*, which is subsequently assigned the address of *employee_1*. This means that the location in memory of the first byte of *employee_1* is stored in *employee_1_ptr*. Each of the remaining statements accesses a member of the data structure indirectly. In each case, note how the indirection operator shows the pointer to be pointing at each member. By comparing this with previous examples, it should be clear that *'employee_1_ptr->'* is doing exactly the same job as *'employee_1'*. To emphasize this, Program 1.8 is a modification of Program 1.7, in which data structure members are accessed indirectly using a pointer rather than being fully qualified.

/* Program 1.8 - Reading and writing a data structure using pointers */

```
#include <stdio.h>

int main(void)
{

struct employee
  {
  int number;
  char family_name[101];
  float salary;
  };

struct employee employee_1, *employee_1_ptr;

employee_1_ptr = &employee_1;

fprintf(stdout, "Enter employee number:");
fscanf(stdin, "%d", &employee_1_ptr->number);
fprintf(stdout, "Enter employee family name:");
fscanf(stdin, "%s", &employee_1_ptr->family_name);
fprintf(stdout, "Enter employee salary:");
fscanf(stdin, "%f", &employee_1.salary);

fprintf(stdout,"Employee: name is %s   number is %d  salary is %f\n",
                           employee_1_ptr->family_name,
                           employee_1_ptr->number,
                           employee_1_ptr->salary);
return(0);
}
```

The differences between Programs 1.7 and 1.8 are, firstly, that the latter contains a pointer, *employee_1_ptr*, of type *struct employee*, which is used to store the address of *employee_1*; and, secondly, that the member variables within *employee_1* are accessed indirectly (pointed to) using *employee_1_ptr->*.

Tutorial 1.15
Implement Program 1.8. Write brief notes on the action of each statement in the program.

Tutorial 1.16
Modify Program 1.8 to allow it to read, store and display data for three employees. The data should be stored in three structures called *employee_1*, *employee_2* and *employee_3*. Declare and use an array of three pointers of type *struct employee* to store the addresses of these structures and to access their members indirectly.

Chapter review

This chapter has concentrated on the different types of data that can be processed in C programs. C specifies a small set of fundamental data types that can be used in declaration statements to create variables. Variables are names that the programmer gives to individual items of data. By using such names, data can be manipulated in a program without the programmer needing to know exactly where they are stored in memory. Various sections in this chapter have demonstrated the rules that need to be followed when creating variables in declaration statements. Other sections have concentrated on forming and using collections of data through the creation and use of arrays and data structures. Arrays are useful but have limitations. To partially overcome these limitations, C allows data structures to be created as programmer-defined data types. Using these, the programmer can design and name variables in ways that reflect the collections of data that need to be processed.

2

Introduction to Executable Statements

2.1 Introduction

Executable statements are those that either process information in some way, for example performing calculations, or use information to control and co-ordinate such processing. Thus, executable statements can be divided into **processing statements** and **control statements**. This chapter will concentrate on the essential features of processing statements. Control statements, such as function calls, decisions and loops will be considered in later chapters.

All executable statements involve the use of **operators** and **operands**. Operands are items of data – variables, constants and values returned from functions. The latter will be discussed in more detail in Chapter 3. Until then, however, simply consider that operands can be variables and constants of types *int, char, float, double*, pointer, elements of arrays and members of data structures. Operators are symbols that define the actions that a computer can perform using various operands. This chapter provides an introduction to those operators that are generally useful in a broad range of programming activities, categorizing them as follows:

- Arithmetic operators (Section 2.2).
- Relational and logical operators (Section 2.3).
- Identifying operators (Section 2.4).
- Miscellaneous (Section 2.5).

Typically, executable statements contain various combinations of the above operator types. Also, executable statements often contain variables and constants of different data types. In such situations, C uses two sets of rules to carry out the required operations. The first, known as **type conversions**, specify how the data types of variables can change temporarily within calculations to prevent loss of accuracy. These are considered in Section 2.5. The second set of rules specifies the **precedence** of operators. In other words, these rules specify the order in which operators are used within a statement. Each operator has a default precedence, which will be considered in Section 2.6.

2.2 Arithmetic operators

The arithmetic operators are:

=	Assignment
+	Addition
–	Subtraction
*	Multiplication
/	Division
%	Modulus
+=	Add, then assign
–=	Subtract, then assign
*=	Multiply, then assign
/=	Divide, then assign
++	Increment
–	Decrement

The assignment operator copies or assigns the value of the operand on its right to the operand on its left, for example:

```
int A, B = 7;
A = B;
```

The previous declaration statement specifies that A and B are variables of type *int* and initializes B. The executable statement then assigns (copies) the value held in B to A. Note that the '=' operator in C is not the same as 'equals' in mathematics. For example,

```
A = A + 1;
```

takes the value held in A, adds 1 to it and copies (assigns) the result back to A.

Generally, the addition, subtraction, multiplication and division operators work in the same way in C as they do in mathematics. However, it should be noted that the subtraction operator can also be used with a single operand. Thus, the statement:

B = –B;

has the effect of changing the sign of variable *B* and then assigning the result back to *B*, equivalent to multiplying *B* by –1.

Next, the modulus operator, %, finds the remainder of the division of two integer values, thus:

```
int A = 6, B = 4, C;
C = A % B;        the value 2 is assigned to variable C
C = B % A;        the value 4 is assigned to variable C
```

It should be noted that the application of an operator to one or more operands, such as *A + B*, gives a result without that result being assigned to some other variable. Exactly what happens to this result depends on the context in which the operation is performed. For example:

```
A * B;        the result is discarded when the next operation is executed
C = A * B;    the result of this operation is stored in C
D = A * B + C; the result of A * B is discarded after being added to the value of
              C; the overall result is then stored in D
```

C additionally provides several 'short hand' operators:

```
int A = 10;
A += 2;    adds 2 to A, storing the result in A (A now holds 12)
A –= 2;    subtracts 2 from A, storing the result in A (A now holds 10)
A *= 5;    multiplies A by 5, storing the result in A (A now holds 50)
A /= 5;    divides A by 5, storing the result in A (A now holds 10)
```

Finally, C provides increment and decrement operators, '++' and '––', respectively, that act on the operand that is either immediately before or after them:

```
int A, B = 5, C = 2;
A = B + C++;  C incremented by 1 after it has been added to B (result: A = 7)
A = B – ++C;  C incremented by 1 before subtracting it from B (result: A = 1)
A = ++B + C;  B incremented by 1 before being added to C (result: A = 10)
A = B – + C;  B decremented by 1 after being added to C (result: A = 10)
```

Program 2.1 demonstrates a simple calculation, using several arithmetic operators.

```
/* Program 2.1 - Demonstration of arithmetic operators   */
/* Calculating the volume of material in a pipe.          */

#include <stdio.h>

int main(void)
{
double  pi = 22.0 / 7.0,
        outer_diameter,
        outer_area,
        inner_diameter,
        inner_area,
        pipe_length,
        pipe_volume,
        pipe_area;

fprintf(stdout, "Enter outer diameter of pipe (m.):");
fscanf(stdin, "%lf", &outer_diameter);
fprintf(stdout, "Enter inner diameter of pipe (m.):");
fscanf(stdin, "%lf", &inner_diameter);
fprintf(stdout, "Enter length of pipe (m.):");
fscanf(stdin, "%lf", &pipe_length);

outer_area = pi * outer_diameter * outer_diameter / 4.0;
inner_area = pi * inner_diameter * inner_diameter / 4.0;
pipe_area = outer_area - inner_area;
pipe_volume = pipe_area * pipe_length;

fprintf(stdout,"\n\nouter_diameter: %le m.\n", outer_diameter);
fprintf(stdout,"inner_diameter: %le m.\n", inner_diameter);
fprintf(stdout,"pipe_length: %le m.\n", pipe_length);
fprintf(stdout,"pipe_volume: %le cubic m.\n", pipe_volume);

return(0);
}
```

Before looking at the detail in Program 2.1, it is useful to look at its 'overall shape'. The executable statements have been divided into three groups – reading required data, processing and outputting results. This has been done to emphasize the distinction between these major tasks that the program has to perform and is a small step towards considering functions in Chapter 3. Prior to the executable statements, all of the required variables have been declared as type *double*. Note how the constant *pi* has been initialized, by assigning to it the result of a calculation. This is a useful trick to ensure that the value of *pi* will be as accurate as the *double* data type will allow. Also note that the name of each variable has a clear meaning within the problem being solved. This is important when writing software to solve 'real' problems because it goes a long way to explaining what the program actually does.

The first group of executable statements prompt for and read the outer and inner diameters of the pipe and its length. Note how the wording of the prompts is consistent with the names of the variables used to store the relevant data. By including *(m.)* in each prompt, the program is also helping the user to be consistent in their use of units for each item of input. The middle group of executable statements carry out the required calculation. It may be surprising, but C does not have a 'squared' operator, hence the calculations for *outer_area* and *inner_area* both involve repetition of the relevant diameters. The final group of statements echo the user's input and also display the calculated result. Again note that the user is shown the units of each displayed value.

Tutorial 2.1
Implement Program 2.1 and make brief notes on its operation. Run the program with various input values.

Tutorial 2.2
Convert Program 2.1 so that all of the variables relating specifically to the pipe are members of a suitably named data structure. Where they are required, each member should be accessed by fully qualifying it. Note that C does not allow a member variable to be initialized when a data structure is being defined as a new data type.

2.3 Relational and logical operators

The C language uses relational operators to make comparisons between operands. The operands of relational operators can be of any data type. The relational operators are:

> greater than
>= greater than or equal to
< less than
<= less than or equal to
== equal to
!= not equal to

The result of applying each of these operators is either TRUE or FALSE, represented in C by numerical values 1 and 0, respectively. For example:

```
int A = 6, B = 2;
A > B;          this relationship is TRUE, numerical value of 1
A < B;          this relationship is FALSE, numerical value of 0
A == B;         this relationship is FALSE, numerical value of 0
A != B;         this relationship is TRUE, numerical value of 1
```

Program 2.2 displays the numerical result of each relational operator, given two integer operands supplied by the user.

```
/* Program 2.2 - Demonstration of relational operators */

#include <stdio.h>

int main(void)
{
int A, B;

fprintf(stdout, "Demonstration of relational operators\n");
fprintf(stdout, "Enter two integer values:");
fscanf(stdin, "%d %d", &A, &B);
fprintf(stdout, "\nA = %d   B = %d\n", A, B);
fprintf(stdout, "\n1 = TRUE, 0 = FALSE\n\n");
fprintf(stdout, "A > B  = %d\n", A > B);
fprintf(stdout, "A >= B = %d\n", A >= B);
fprintf(stdout, "A < B  = %d\n", A < B);
fprintf(stdout, "A <= B = %d\n", A <= B);
```

```
fprintf(stdout, "A == B = %d\n", A == B);
fprintf(stdout, "A != B = %d\n", A != B);

return(0);
}
```

In Program 2.2 the values of two integers are read from the keyboard and echoed on the screen. These integers are then compared using each relational operator. Since each comparison appears as an argument in a call to the *fprintf* function, the result obtained from each comparison is passed as an argument to *fprintf*. Relational operators are mostly used in decision making and further details of their use and effects will be put aside until Chapter 4.

Tutorial 2.3
Implement Program 2.2 and make brief notes on its operation. Run the program with various input values.

Tutorial 2.4
Convert Program 2.2 to read variables of type *double* and run the program with various input values.

Logical operators are used to group two or more relational operations together in various ways. The logical operators are:

&& meaning logical AND
|| meaning logical OR

As with the relational operators, the AND and OR operators give the results 1 if the relationship is TRUE or 0 if the relationship is FALSE. The result of the AND operator is TRUE if both of its operands are TRUE, and FALSE if either of its operands are FALSE. The result of the OR operator is TRUE if either of its operands are TRUE, and FALSE if both of its operands are FALSE. The use of these operators is demonstrated in Program 2.3 where, given two integers, the program determines if either or both of their sum and difference are greater than zero.

```
/* Program 2.3 - Demonstration of logical operators */

#include <stdio.h>

int main(void)
{
int A, B, C;

fprintf(stdout, "Demonstration of logical operators\n");
fprintf(stdout, "Enter two integer values:");
fscanf(stdin, "%d %d", &A, &B);
fprintf(stdout, "\nA = %d   B = %d\n", A, B);
fprintf(stdout, "\n1 = TRUE, 0 = FALSE\n\n");

C = A + B > 0 && A - B > 0;
fprintf(stdout, "sum AND difference of A and B greater than zero ?: %d\n", C);

C = A + B > 0 || A - B > 0;
fprintf(stdout, "sum OR  difference of A and B greater than zero ?: %d\n", C);

return(0);
}
```

The above program contains two statements in which the integers supplied are compared, both storing the comparison result in variable C. The first of these uses the AND operator with operands that are both results of relational comparisons. If $A + B > 0$ is TRUE, this operand of the AND operator will be TRUE (value 1). Similarly, if $A - B > 0$ is TRUE then the second AND operand will also be TRUE (value 1). If both operands are TRUE, the result of the AND operator will be TRUE and a value of 1 will be assigned to C. Conversely, if either of the operands is FALSE, a value of 0 will be stored in C.

The second comparison statement uses the OR operator with the same operands as before. In this statement, the OR operator will be TRUE, with a value of 1 assigned to C, if either operand is TRUE. The OR operator will be FALSE, with a value of 0 assigned to C, only if both of its operands are FALSE.

Tutorial 2.5
Implement Program 2.3 and make brief notes on its operation.
Run the program with various input values to give all possible
outputs.

Tutorial 2.6
Modify Program 2.3 so that variables *A* and *B* are of type
double. Run the program with various input values to give all
possible outputs.

2.4 Identifying operators

Four operators are frequently used in C programs to identify things:

[ ]   identifies an element of an array
.   fully qualifies a member of a data structure
–>   provides indirect access to members in a data structure whose
      address is stored in a pointer
( )   identifies the **precedence** of operations (also see Section 2.6)

All of the identifying operators in the above list, except for the last,
have already been introduced in Chapter 1. However, several
examples are given below to reinforce their actions.

Firstly, the *[...]* identifier could be called the 'element operator'
which is used to identify a particular element of an array, as in:

```
int A, B[3];
A = B[2];   assigns the value of the third element of array B to A
```

Secondly, the 'dot operator', . , is used with the name of a data
structure to access its member variables, as in:

```
struct PAIR                        Declares a template for a data
                                   structure
    {                              (data type) called struct PAIR
    int A;
    double B;
    };
```

struct PAIR PAIR_1, PAIR_2;	Declares two structures of type struct PAIR.
int A = 2, B = 9;	Declares and initializes *A* and *B*.
PAIR_1.A = 5;	Assigns the value 5 to *A* in *PAIR_1*.
PAIR_1.B = 5;	Assigns the value 5 to *B* in *PAIR_1*.
PAIR_2.A = 10;	Assigns the value 10 to *A* in *PAIR_2*.
PAIR_2.B = 5;	Assigns the value 5 to *B* in *PAIR_2*.
PAIR_1.B = A;	Assigns the value 2 to *B* in *PAIR_1*.
B = PAIR_1.A;	Assigns the value 5 to *B*.
PAIR_1.B = PAIR_2.A;	Assigns the value 10 to *B* in *PAIR_1*.
*PAIR_2.B = (PAIR_1.A + PAIR_2.A) * A;*	Assigns the value 30 to *B* in *PAIR_2*.

In the above example there are three variables called *A* and another three called *B*. In addition to the declaration, *int A = 2, B = 9;*, two structures, *PAIR_1* and *PAIR_2*, are declared, each containing member variables called *A* and *B*. To access any of the member variables in the above example, they have been fully qualified using the dot operator.

Thirdly, as an alternative to fully qualifying the members of a data structure, the indirection operator '->' can be used if the address of the structure has been stored in a pointer. This is demonstrated below by recoding the previous example.

struct PAIR	Declares a template for a data structure
{	(data type), called *struct PAIR*.
int A;	
double B;	
};	

struct PAIR PAIR_1, PAIR_2;	Declares two structures of type *struct PAIR.*
struct PAIR *PAIR_1_ptr, *PAIR_2_ptr;	Declares two pointers of type *struct PAIR.*
int A = 2, B = 9;	Declares and initializes *A* and *B.*
PAIR_1_ptr =&PAIR_1;	Stores addresses of data structures.
PAIR_2_ptr =&PAIR_2;	
PAIR_1_ptr->A = 5;	Assigns the value 5 to *A* in *PAIR_1.*
PAIR_1_ptr->B = 5;	Assigns the value 5 to *B* in *PAIR_1.*
PAIR_2_ptr->A = 10;	Assigns the value 10 to *A* in *PAIR_2.*
PAIR_2_ptr->B = 5;	Assigns the value 5 to *B* in *PAIR_2.*
PAIR_1_ptr->B = A;	Assigns the value 2 to *B* in *PAIR_1.*
B = PAIR_1_ptr->A;	Assigns the value 5 to *B.*
PAIR_1_ptr->B = PAIR_2_ptr->A;	Assigns the value 10 to *B* in *PAIR_1.*
PAIR_2_ptr->B = *(PAIR_1_ptr->A + PAIR_2_ptr->A) * A;*	Assigns the value 30 to *B* in *PAIR_2.*

The final identifying operator in the above list, the precedence operator, (...), is new and is used to fix the sequence in which other operators are used in an executable statement. For example, given:

double A = 1.0, B = 3.0;	
double C = 5.0, D = 8.0, E;	
E = A + B * C + D;	assigns the value 24.0 to *E*, whereas
E = (A + B) * (C + D);	assigns the value 52.0 to *E*

Here, the two executable statements involve the same arithmetic operators and operands, but give different results. In the first of the two statements, the multiplication operator takes default precedence over the addition operators. In the latter statement, the addition operations have been placed within precedence operators, forcing them to take precedence over multiplication.

Tutorial 2.7
Implement the dot operator example, above, as a working program. Add a call to *fprintf* after each executable statement to display ALL of the variables and ensure that the displayed values are consistent with those given in the example. Ensure that members of data structures are accessed only by fully qualifying them.

Tutorial 2.8
Implement the indirection operator example, above, as a working program. Add a call to *fprintf* after each executable statement to display ALL of the variables and ensure that the displayed values are consistent with those from the previous problem. Ensure that members of data structures are accessed indirectly.

2.5 Miscellaneous operators

Three operators are considered here. The first is the '**contents of**' operator, previously mentioned in Section 1.5, in the context of the pointer data type. The second is the *sizeof* operator, used in Chapter 1, Question 1 of the typical examination questions at the end of the book. The final operator is **cast**, which is used to convert between different data types.

The 'contents of' operator uses the * symbol and care must be taken not to confuse it with the multiplication operator. The 'contents of' operator is used to obtain the value stored in a variable when the address of that variable is stored in a pointer. For example:

```
double A, B = 1.0, C = 3.0;
double *B_ptr, *C_ptr;
A_ptr = &A;
```

```
B_ptr = &B;
C_ptr = &C;
A = *B_ptr + *C_ptr;
*A_ptr = *B_ptr + *C_ptr;
```

The last two statements, above, achieve the same result. In each case, the 'contents of' operator is used with pointers that hold the addresses of *B* and *C*, to retrieve the values that *B* and *C* were given when they were declared. In the first instance, the value 4.0 is assigned to variable *A*. In the second instance, **A_ptr* means 'store the value 4.0 as the contents of the variable whose address is held in *A_ptr*'. Significant use will be made of this operator in the next chapter.

The *sizeof* operator is most frequently used with the *fgets* function, discussed in Chapter 5, and dynamic memory allocation, considered in Chapter 6. When *sizeof* is given a data type, it returns the number of bytes needed by that type, for example:

int A;	
A = sizeof(double);	*A* is assigned the value 8 (8 bytes needed to store a variable of type *double*).
struct collection	Defines a new data type.
{	
double X;	Uses 8 bytes.
int Y;	Uses 2 bytes.
float Z[3];	Uses 12 bytes.
};	
struct collection B;	Declares a variable, *B*, of type *struct collection*.
int A;	
A = sizeof(struct collection);	*A* is assigned the value 22 (22 bytes needed to store a variable of type *struct collection*).

The cast operator allows the programmer to break the rules that C normally applies to the specification of data types. When an operator requires two or more operands or when an executable statement involves several operators and operands, it is often the case that the operands are of different data types. When this happens, the operands currently being processed are automatically converted to a common data type, through type conversion, before an operator is applied. This is done to preserve the accuracy of calculations. In general terms, the two most basic automatic conversion rules are

> When *int* occurs with *float* or *double*, the *int* operand is temporarily converted to a *float* or *double*, as required.

For example:

```
float A, B;
int I;
A = B + I;   (I is converted to a float before the + operator is applied)
```

Similarly, when operands of types *double* and *float* appear together, the *float* is temporarily converted to a *double*:

```
double A, B;
float C;
A = B + C;   (C is converted to a double before the + operator is applied)
```

There are, however, many situations where programming mistakes can be made by using operands of mixed data type. For example, the following are all valid C statements, but may not give the results that the programmer intended:

```
float A = 10.7;
int B, C = 5;
B = A + C;
```

Here, variable C is promoted to *float* so that the result of the + operator is a *float*. However, since the target variable, B, is an integer, the = operator discards the fractional part of $A + C$ and only copies the integer part to B. Thus, in this example B is assigned the value 15, rather than 15.7. Also:

```
float D, E;
double F;
D = E + F;
```

This second example is more subtle, in that it involves variables of similar data type, but different precision. Here, E is promoted to a *double* so that the result of the + operator is a *double*. The result of this is then stored in a variable of type *float*.

Two problems may arise here. Firstly, if the result of $E + F$ is greater than the maximum value that can be stored in a variable of type *float* this will cause the value of D to be corrupted. C handles this in a controlled way rather than terminating the program. However, D will contain a value that is unusable in subsequent statements. Secondly, even if the result of $E + F$ is not too large to be stored in a *float*, information will be lost because of the difference in precision of *float* and *double* type variables (see Section 1.4). Similar problems can occur when mixing variables of type *int*, *unsigned int*, *short int*, etc. One obvious, if inelegant, way to avoid such problems is to always use the *long int* data type for integer variables and *double* data type for real number variables. However, this approach may cause its own problems by at least wasting, and possibly running out of memory. When such problems are anticipated, the cast operator can be used to force a change of data type as follows:

```
target_variable = (data type)source_variable;
```

where *(data type)*, is the cast operator, used to convert the data type of *source_variable* to that of *target_variable*. For example:

```
float D, E;
double F;
D = (float)(E + F);
```

would overcome the 'precision' error in the previous example by rounding up prior to assignment. Note how the + operation has been enclosed within brackets, forcing it to take precedence over the cast operator. It is worth noting, however, that the cast operator cannot fix the problem of trying to store too large a value in a variable, for example when the value of $(E + F)$, above, is beyond the range of values that can be stored in D.

2.6 Operator precedence

The precedence, or importance, of an operator indicates its priority when it occurs in a statement along with other operators. An operator having higher precedence will be carried out before an operator of lower precedence. Table 2.1 lists all of the operators in C (some not discussed in this book) in order of decreasing default precedence. Operators on the same line have equal precedence. Several points should be noted about Table 2.1:

- Where ( ) occurs in function calls, next chapter, and in nested operations within a statement, function calls take precedence.
- The highest occurrence of + and – are unary operators.
- The highest occurrence of * is the 'contents of' operator.
- The highest occurrence of '&' is the 'address of' operator.

Table 2.1 *Operators in decreasing order of precedence*

[ ] . ( ) –>
* – + ++ — & | sizeof (data type)
% * /
+ –
>> <<
> >= < <=
!= ==
&
^
|
&&
||
?:
= –= += /= *= %= >>= <<= |= &= ^ =
,

The second example in Section 2.4, repeated below, provides a good example of operator precedence:

```
double A = 1.0, B = 3.0;
double C 5.0, D = 8.0, E;
E = A + B * C + D;        assigns the value 24.0 to E, whereas
E = (A + B) * (C + D);    assigns the value 52.0 to E
```

Referring to Table 2.1, the multiply operator, *, is found in line 3 and the addition operator, +, is found in line 4 hence the latter has lower precedence. In the first assignment of a value to *E*, above, the multiplication operator is executed before either of the addition operators. If, as in the second assignment of a value to *E*, we want to perform the additions before the multiplication, we must enclose the addition operators and their operands within brackets. From Table 2.1, '( )' has a higher precedence than either * or +, forcing the result of each bracketed operation to become an operand for the multiplication operator.

A second example of precedence occurs in Program 2.3. The relevant program statements are repeated below:

```
C = A + B > 0 && A - B > 0;
fprintf(stdout," sum AND difference of A and B greater than zero ?: %d\n", C);
C = A + B > 0 || A - B > 0;
fprintf(stdout," sum OR difference of A and B greater than zero ?: %d\n", C);
```

In each of the assignment statements, above, the addition and subtraction operators (line 4 of Table 2.1) are used first. The '>' and '<' operators (line 6 of Table 2.1) are executed next. Finally the *&&* and || operators (lines 11 and 12 of Table 2.1) are then applied.

Chapter review

This chapter has introduced executable statements by considering several C operators and the various ways in which they can be combined to perform useful tasks. The decision to exclude certain operators from this chapter has been made on their relatively specialized applications. It is useful to recognize that operators in C can be divided into several classes, with operators in any particular class providing a distinct aspect of C's functionality. This means that it has been quite easy in this chapter to explore the Arithmetic and Identifying operators, but more detailed consideration of Relational and Logical operators must wait until their more typical use in decision making, in Chapter 4 onwards.

C provides a broad range of arithmetic operators that can be combined in many ways within executable statements. Care should be taken, however, over the data types of their operands and of the results that they produce, to ensure that problems don't arise with truncation, loss of precision and attempts to store values that are outside the possible range that a variable can hold. The 'short hand' operators are very convenient when you remember that they exist, but they are only available to combine the elemental + – * / operations with assignment of their result. Particular caution needs to be exercised over the increment and decrement operators and whether they appear before or after the variable that they are operating on.

The typical examination questions for this chapter at the end of the book are intended to develop some proficiency in using some of the operators considered here in the context of several engineering and science related problems. An additional aim in doing this is to provide practice in writing whole (although very small) programs that do something useful. After all, it is hard to think of other reasons for writing software.

3

Introduction to Functions

3.1 Introduction

Referring back to the Introduction, all C programs contain at least one function, called *main*. When we design a C program whose overall task can be divided into a collection of smaller sub-tasks, we usually build the program by creating a function to perform each sub-task. There are several reasons why this is a good idea:

- To design a program we often use some method of **software engineering**. Each approach to software engineering divides the required task into sub-tasks, modules, sub-systems or processes of various types. Functions are a natural way of implementing such designs in C.
- Even without software engineering, functions allow the structure of the program to reflect the structure of its application.
- Using functions removes the need to repeat identical groups of statements within programs when the same task must be performed several times.
- The use of functions allows libraries of frequently used software to be built up and re-used in different programs.
- C functions can be used as operands in executable statements, allowing the creation of compact and efficient programs.

Functions are used by **calling** them from other functions. When a function is used, it is referred to as the '**called function**'. Such functions often use data that is passed to them from the calling function. Data is passed from a **calling function** to a called function by specifying the names of variables in an **argument list**. It is important to remember that *argument lists only pass data from a calling function to a* *called function*. An argument list cannot be used to pass or return data from a called to a calling function. This is because, whenever an argument list is used, the called function always makes a copy of each variable that is supplied to it by the calling function. The called function then performs its operations using the copies.

Data can be passed to functions via an argument list in two ways, **passing by value** and **passing by reference**:

- **Passing by value:** the data value stored in a variable is passed. For example, suppose that a variable, A, is used in a function and has a value of 1.6. If the function calls another function and A appears in the argument list, then A is being passed by value. When this is done, the called function creates a new variable, say B, and copies the value of A into it. If the called function subsequently changes the value of B from 1.6 to 3.2, only the value of B is modified, variable A in the calling function still contains the value 1.6.

- **Passing by reference:** the address (the location in memory) of a variable is passed. For example, again suppose that a variable, A, is used in a function and has a value of 1.6. As discussed in Chapter 1, the address of A can be found using $\&A$. This address can also be stored in a pointer variable, say A_ptr. Suppose that the function calls another function. If $\&A$, or A_ptr, appears in the argument list for the called function, then A is being passed by reference. The address of A is being given to the called function. When this happens, the called function creates a new pointer variable, say B_ptr, and copies the passed address into it. Thus, both the calling function and the called function now know the address of variable A. The called function can now use the 'contents of' operator with B_ptr, e.g. $*B\_ptr = 3.2;$ to change the value stored at the memory location pointed to by B_ptr. This also changes the value of A in the calling function, from 1.6 to 3.2.

Passing by value and passing by reference seem to be quite different methods of giving data to a function. However, they are, in fact, exactly the same:

- In passing by value, the called function makes a copy of a variable that contains data.
- In passing by reference, the called function makes a copy of a variable that contains an address.

When a called function is intended to simply use the data that is passed to it, it is best practice to pass the data by value. If the called function is intended to change the data passed to it, the data must be passed by reference.

A called function can always pass data back to the calling function through the *return* statement. In fact, if data is passed to the called function by value, this is the only way that data can be passed back from the calling function. In contrast, when data is passed to a function by reference, data can also be passed back by changing the values of the referred variables.

Individual variables having the data types mentioned in Sections 1.2 to 1.5, inclusive, can always be passed back to the calling function through the *return* statement. In addition, arrays, character strings and data structures, Sections 1.6, 1.7 and 1.8, respectively, can be passed back via the *return* statement in certain versions of C.

After describing the essential statements needed in any function, the remainder of this chapter is concerned with the interface between calling and called functions and ways in which variables of different data types can be passed between them.

3.2 Essential statements in any function

Every C function must contain statements of the following types:

```
returned_data_type function_name(argument_list)
{
declaration statements
executable statements
return(variable of returned_data_type)
}
```

Every function can pass back information of some kind to the function that called it. Hence, in the first line, above, *returned_data_type* represents any of the C data types, *char, float, double,* all of the different *int* types and pointers that were discussed in Chapter 1. Also, *returned_data_type* can be '*void*', meaning that no data is returned from the called function. In the first line *returned_data_type* is followed by the name of the function, which must be unique within the program. A function may or may not have an argument list, depending on whether the called function needs to receive data from the calli function. Remember that, where an argument list is used, it only a data to be passed from the calling to the called function. It is not possible to pass data back from the called to the calling function via an argument list. If no arguments are passed to a function, the argument list is specified as '*(void)*'. Following the argument list, the opening bracket, {, marks the start of the statements within the function. This bracket is matched by the closing bracket, }, after the last executable statement in the function. The final executable statement in the function should be a *return* statement. The *return* statement can have an argument list that contains the name of just one variable, whose value has been set inside the called function. If a variable is specified in a *return* statement, its data type must be the same as *returned_data_type* in the first line of the function.

All of the complete programs considered in Chapters 1 and 2 are specific examples of these essential statements. The general scheme that has been used is:

```
int main(void)
{
Declaration statements
Executable statements
return(0);
}
```

Remember that *main* is a C function and, consequently, must adhere to the general rules for all C functions. The unique thing about *main* is that it is always the first function to be used in a C program, hence we can think of the operating system as the calling function and *main* as the called function. Following the previously discussed rules, the argument list for *main* in previous chapters is *void*, meaning that it receives no data from the operating system. However, *main* is intended to pass an integer value back to the operating system through the *return* statement. In all cases considered so far, this value is zero. The intention in structuring previous examples in this way has been to provide a skeleton function structure that should, by now, be quite familiar, easily recognized and, hopefully, easily changed to account for the new material in this chapter.

between calling and called functions

used by another function, the calling function
tion prototype statement. Essentially, a function

prototype is a declaration statement for a function that the calling function will use. There must be a separate function prototype statement in the calling function for every function that it uses. In addition, the name of each called function must appear in an executable statement, where it is required to perform its task.

The most simple situation occurs when the calling function passes no data to the called function and no data is returned. This would be typical of using a function to display a message, as shown in Program 3.1.

```
/* Program 3.1 - Calling a function with no argument list and no returned  */
/* data                                                                    */

#include <stdio.h>

int main(void)
{
void print_message(void);        /* function prototype      */

print_message();                 /* call the function       */
return(0);
}
```

```
/* Function: print_message */

void print_message(void)
{
fprintf(stdout,"This message has been displayed"
            "using the print_message function"
            "in program 3.1\n");
return;
}
```

Program 3.1 consists of two functions, one is called *main* and the other is called *print_message*. *main* contains a function prototype (a function declaration statement) and two executable statements. The function prototype indicates that *main* will call a function called *print_message*. The argument list for *print_message* is *(void)*, meaning that no data will be passed to the function from *main*. Also, the function prototype shows that *print_message* does not return any data, because the *returned_data_type*, in front of the function name, is also *void*.

The first executable statement in *main* calls the *print_message* function. When this statement is executed, the *print_message* function takes over from *main*, which waits until all of the statements in *print_message* have been executed.

Looking at the *print_message* function, the first line after the comments specifies the name of the function, along with its argument list and the data type of the returned variable. Compare this line with the function prototype and the statement that uses *print_message*, in *main*. All of these statements are consistent: the function to be used by *main* is called *print_message* and no data is returned from the function and no data is passed to it. If these statements are not consistent, the program will not work correctly.

Looking inside the *print_message* function, the first executable statement is a call to the *fprintf* function, which displays a message on the screen. *fprintf* is a standard C function and its function prototype statement is stored inside a library, called *stdio.h*. This library is included in any C program by using the *#include <stdio.h>* statement.

Tutorial 3.1
Implement Program 3.1 and make brief notes on its operation.

3.4 Non-empty argument lists and return statements

This section contains five example programs, each consisting of two functions. The rather trivial objective of these programs is to add two numbers together and display their sum on the screen. One of the functions, *main*, specifies the values that are to be added together and then passes them to the second function, called *add_numbers*. The *add_numbers* function adds the values and passes their sum back to *main*, which then displays the result on the screen. The difference between the programs is that data is passed between *main* and *add_numbers* in different ways, as follows:

- Program 3.2: Data passed by value to *add_numbers*, result passed back to *main* through the *return* statement.
- Program 3.3: Data passed by reference to *add_numbers*, result passed back to *main* through the *return* statement.

- Program 3.4: Data passed by reference to *add_numbers*, result passed back by reference from *add_numbers* to *main* through a third argument.
- Program 3.5: Data given to *add_numbers* stored in a data structure, passed by value. Result passed back to *main* through the *return* statement.
- Program 3.6: Data going to *add_numbers* stored in a data structure, passed by reference. Result passed back to *main* through a member of the data structure.

Program 3.2: Data passed by value to *add_numbers*, result passed back to *main* through the *return* statement.

```
/* Program 3.2 */

#include <stdio.h>

int main(void)
{
float A = 6.0, B = 10.0, C;
float add_numbers (float, float);        /* function prototype      */

C = add_numbers (A, B);                  /* call function           */

fprintf(stdout," The sum of %f and %f is %f \n", A, B, C);
return(0);
}


/* Function: add_numbers */

float add_numbers (float X, float Y)
{
float D;

D = X + Y;
return(D);
}
```

The first declaration statement in *main* declares three *float* type variables. The first two variables are initialized and will be passed to *add_numbers*. The third is be used to store the value passed back from *add_numbers*.

The function prototype statement in *main* specifies that it will use a function called *add_numbers*. It also specifies, firstly, that the called function expects two *float* values to be passed to it and, secondly, that *add_numbers* will pass back a *float* type value via its *return* statement. Note that the argument list in the function prototype only specifies the data types to be passed to *add_numbers*, rather than the actual variables.

The first executable statement in *main* calls the *add_numbers* function. It is useful to read this from right to left:

> The values of *A* and *B* are passed to *add_numbers*, which returns a value that is stored in *C*.

It is important to recognize that calling *add_numbers* and storing the value that it returns are two distinct operations. In effect, when *add_numbers* returns its value, the value replaces the function in the calling statement and is then stored in *C*.

Looking at the *add_numbers* function, notice that two variables, *X* and *Y*, are declared in its argument list. These are the variables that the values of *A* and *B* are copied into. It is essential that the sequence of the variables and their data types are identical in the argument lists of the calling and called functions. To do its job, *add_numbers* adds the values of *X* and *Y* together and assigns the result to *D*. Since *D* is specified in the *return* statement, its value is passed back to the calling function. Note that the data type of the variable returned from *add_numbers* is exactly the same as the data type of the variable appearing in the *return* statement.

Tutorial 3.2
Implement Program 3.2 and make brief notes on its operation.

Program 3.3: Data passed by reference to *add_numbers*, result passed back to *main* through the *return* statement.

```
/* Program 3.3 */

#include <stdio.h>

int main(void)
{
float A = 6.0, B = 10.0, C;
float *A_ptr, *B_ptr;
float add_numbers (float *, float *);          /* function prototype        */

A_ptr = &A;
B_ptr = &B;

C = add_numbers (A_ptr, B_ptr);                /* call the function         */

fprintf(stdout, " The sum of %f and %f is %f \n", A, B, C);
return(0);
}



/* Function: add_numbers */

float add_numbers (float *ptr1, float *ptr2)
{
float D;

D = *ptr1 + *ptr2;
return(D);
}
```

In Program 3.3, two pointers of type *float* are declared in *main*, which are used to store the addresses of variables *A* and *B*. Also, the argument list in the prototype statement for *add_numbers* indicates that two pointers of type *float* must be passed to the function. This is consistent with the first line of the *add_numbers* function, where the argument list declares two pointers, *ptr1* and *ptr2*, that will store copies of the data passed to it.

When *main* calls *add_numbers*, the value of *A_ptr* is copied to *ptr1* and the value of *B_ptr* is copied to *ptr2*. So, in this program, the addresses of *A* and *B* are passed, rather than the values of the data stored in *A* and *B*. Thus, *A* and *B* are passed by reference. The *add_numbers* function accesses the values of *A* and *B* using the 'contents of' operator, *. It then adds these values together and stores the result in *D*. Finally, the value of *D* is passed back to *main*, via the *return* statement. This means that, in *main*, the returned value replaces the call to *add_numbers* and is then stored in *C*.

Tutorial 3.3
Implement Program 3.3 and make brief notes on its operation.

Program 3.4: Data passed by reference to *add_numbers*, result passed by reference from *add_numbers* to *main* through a third argument.

```
/* Program 3.4 */

#include <stdio.h>

int main(void)
{
float A = 6.0, B = 10.0, C;
float *A_ptr, *B_ptr, *C_ptr;
void add_numbers (float *, float *, float *);        /* function prototype      */

A_ptr = &A;
B_ptr = &B;
C_ptr = &C;
add_numbers (A_ptr, B_ptr, C_ptr);                   /* call the function       */

fprintf(stdout, "The sum of %f and %f is %f \n", A, B, C);
return(0);
}
```

/* Function: add_numbers */

```
void add_numbers (float *ptr1, float *ptr2, float *ptr3)
{

*ptr3 = *ptr1 + *ptr2;
return;
}
```

Looking at the function prototype statement for *add_numbers* in *main*, no data is passed back from *add_numbers* to *main* through the *return* statement. Instead, the address of a third variable, *C*, is passed to the *add_numbers* function. When the *add_numbers* function is called, the values of *A_ptr*, *B_ptr* and *C_ptr* are copied to *ptr1*, *ptr2* and *ptr3*, respectively. In the first executable statement within *add_numbers*, the 'contents of' operator, *, is used with *ptr1* and *ptr2* to retrieve the values stored in *A* and *B*. The sum of these values is then stored in *C* by using the 'contents of' operator, *, with *ptr3*. As a consequence of this, the value of the variable *C* in *main* has been changed from inside *add_numbers*.

Tutorial 3.4
Implement Program 3.4 and make brief notes on its operation.

Program 3.5: Data given to *add_numbers* is stored in a data structure, passed by value and result passed back to *main* through the *return* statement.

/* Program 3.5 */

```
#include <stdio.h>

struct numbers
  {
  float A;
  float B;
  };
```

```
int main(void)
{
struct numbers set_1;
float C;
float add_numbers (struct numbers);        /* function prototype        */

set_1.A = 6.0;
set_1.B = 10.0;

C = add_numbers (set_1);                    /* call the function        */

fprintf(stdout, "The sum of %f and %f is %f \n", set_1.A, set_1.B, C);
return(0);
}


/* Function: add_numbers */

float add_numbers (struct numbers set_2)
{
float D;

D = set_2.A + set_2.B;
return(D);
}
```

The first point to note about Program 3.5 is that a data structure, called *struct numbers* is defined outside of both functions. It is important to understand that defining *struct numbers* outside of both functions only provides a common definition of the *struct numbers* data type. It does not mean that data is shared by each function.

Looking at *main*, the first declaration statement creates a variable, called *set_1*, of data type *struct numbers*. In addition to this, the *add_numbers* function prototype statement specifies that a variable of type *struct numbers* will be passed by value to *add_numbers*. The function prototype also specifies that *add_numbers* will return a value of type *float*.

After the function prototype statement, values are assigned to the members, *A* and *B*, of *set_1*. The *add_numbers* function is then called and *set_1* is passed by value. Looking at the *add_numbers* function, its argument list declares *set_2* as a *struct numbers* type variable to hold a copy of the data in *set_1*. To create this new variable, *add_numbers* must use the correct data type. This is why the data type *struct numbers* is defined outside of each function.

Inside *add_numbers*, in order to add the values of members *A* and *B* in *set_2*, each must be fully qualified, as shown. Having stored the sum of members *A* and *B* in *D*, the value of *D* is then passed back to *main* through the *return* statement.

Tutorial 3.5
Implement Program 3.5 and make brief notes on its operation.

Program 3.6: Data going to *add_numbers* is stored in a data structure, passed by reference and result passed back to *main* through a member of the data structure.

```
/* Program 3.6 */

#include <stdio.h>

struct numbers
  {
  float A;
  float B;
  float C;
  };

int main(void)
{
struct numbers set, *set_ptr;
void add_numbers (struct numbers *);          /* function prototype       */
set_ptr = &set;
set_ptr->A = 6.0;
set_ptr->B = 10.0;

add_numbers (set_ptr);                          /* call the function        */
fprintf(stdout, " The sum of %f and %f is %f \n",      set_ptr->A,
                                                       set_ptr->B,
                                                       set_ptr->C);
return(0);
}
```

/* Function: add_numbers */

```
void add_numbers (struct numbers *new_set_ptr)
{
new_set_ptr->C = new_set_ptr->A + new_set_ptr->B;
return;
}
```

Program 3.6 is similar to Program 3.5 in that *struct numbers* is again defined outside of each function so that each has a common understanding of this data type. In *main*, a variable, called *set* and a pointer, *set_ptr*, are declared to be of data type *struct numbers*. Following this, the function prototype statement specifies that a pointer of type *struct numbers* will be passed to *add_numbers*, and that *add_numbers* will not return any data via the *return* statement. Next, the address of *set* is stored in *set_ptr* and numerical values are assigned to the members, *A* and *B*, of *set*. The *add_numbers* function is then called, passing the value of *set_ptr* to it.

When *add_numbers* is called, it creates a pointer, called *new_set_ptr*, and copies the value of *set_ptr* into it. Having done this, both *set_ptr*, in *main*, and *new_set_ptr*, in *add_numbers*, hold the address of *set*, so its members can be accessed and changed inside either function. Looking at the first executable statement in *add_numbers*, *new_set_ptr* is used to obtain the values of members *A* and *B*, and to store their sum in member *C*. As a result of this, when the *add_numbers* function has finished, the new value of member *C* is available inside *main*.

Tutorial 3.6
Implement Program 3.6 and make brief notes on its operation.

3.5 Using functions to read and write data

In general terms, the instructions in most programs can be divided into three groups: reading data into the program (e.g. from the keyboard or a file), processing the input data in some way to obtain a result (e.g. performing some calculations), and transferring the result out of the program (e.g. to the screen or a file). These actions can be thought of as input, processing and output.

For all but the most simple programs it good practice to implement separate functions that are dedicated to each of these tasks. In such programs, *main* would call an input function, a processing function and an output function. Programs 3.7 and 3.8, in this section, aim to demonstrate this approach in a relatively simple way, by building separate functions, called by *main*, for getting data into and out of a program. An example of the more general case, involving an additional function for the processing step is dealt with in Section 3.6.

Program 3.7: Passing individual variables

The objective of Program 3.7 is to read an integer, a decimal and a character string from the keyboard and to display them on the screen. Looking at the listing of the program, note that it contains three functions: *main*, *read_data* and *write_data*.

```
/* Program 3.7 - Use of functions for input and output of data */

#include <stdio.h>

int main(void)
{
int a;
float b;
char c[11];

void read_data(int *, float *, char *);        /* function prototype */
void write_data(int, float, char []);          /* function prototype */

read_data(&a, &b, c);
write_data(a, b, c);

return(0);
}


/* Function: read_data - reads an int, float and char string. */

void read_data(int *int_ptr, float *float_ptr, char *char_ptr)
{
```

```
fprintf(stdout,"Supply an integer, a float and a string (max. 10 chars):");
fscanf(stdin,"%d %f %s", int_ptr, float_ptr, char_ptr);
return;
}
```

/* Function: write_data - displays an int, float and char string. */

```
void write_data(int i, float j, char k[])
{
fprintf(stdout,"The supplied data is:  %d %f %s\n",i, j, k);
return;
}
```

Considering *main* in Program 3.7, three variables, *a*, *b*, *c*, are declared to hold the data that will be read and displayed. Following these, *main* contains a prototype statement for each of the functions, *read_data* and *write_data*, that it will use. Note that the argument list for *read_data* specifies that three pointers will be passed to it. This is important because *read_data* is intended to get values from the user and store them somewhere. The argument list for *read_data* says that three variables will be passed by reference, so that *read_data* can use them to store the user's input. The prototype statement for *write_data* specifies that variables of type *int*, *float* and a character string will be passed to it by value. Passing by value is appropriate here because the *write_data* function is intended to simply use the values passed to it, rather than change them.

Consistent with the prototype statements, *main* first calls *read_data*, giving it the addresses of the previously declared variables, then calls *write_data* giving it the values of the same variables. By passing three empty variables to *read_data*, main is giving it three empty 'containers' that *read_data* can put the user's data into. In the call to *read_data* it may appear that the character string, *c*, is passed by value. However, this is a quirk of C, because the name of an array is also a pointer to its address, so the array name is used in passing by value and in passing by reference. Inside the *read_data* function, its argument list declares three pointers of the appropriate data types to hold its copies of the pointers supplied by *main*. These copies are then given to *fscanf* so that the user's inputs can be stored in memory. Having done this, control then passes back to *main*.

When *main* calls *write_data*, the values of *a*, *b* and *c* are copied into *i*, *j* and *k*, which have been declared in the *write_data* argument list. These copies are then passed on to *fprintf* to be displayed.

In this example *main* has been used to co-ordinate the actions of other functions. Although the overall action of the program is trivial and rather contrived (the same task could have been done simply by calling *fscanf* and *fprintf* directly from *main*) the organization of the program into specific functions for specific tasks is very important in building software for any size of problem.

> Tutorial 3.7
> Implement Program 3.7 and make brief notes on its operation.

Program 3.8: Passing a data structure by reference

Program 3.8 does the same job as Program 3.7 but, this time, the three variables are grouped together in a data structure, called *set_1*. The data type for *set_1* is the data structure, *struct set*, which is defined outside of each function. Since *read_data* is intended to change the values of members in *set_1*, *main* must pass *set_1* by reference to *read_data*. This is done using pointer *set_1_ptr*. In contrast, *write_data* is intended to display the values of the member variables, so *set_1* is passed by value from *main* to *write_data*.

```
/* Program 3.8 - Use of functions and passing a data structure by    */
/* reference                                                          */

#include <stdio.h>

struct set
  {
  int a;
  float b;
  char c[11];
  };

int main(void)
  {
  struct set set_1;
```

```
struct set *set_1_ptr;
void read_data(struct set *);
void write_data(struct set);

set_1_ptr = &set_1;
read_data(set_1_ptr);
write_data(set_1);
return(0);
}
```

/* Function: read_data - reads an int, float and char string */

```
void read_data(struct set *set_2_ptr)
{
fprintf(stdout,"Supply an int, float and character string (max. 10 chars):");
fscanf(stdin,"%d %f %s", &set_2_ptr->a, &set_2_ptr->b, set_2_ptr->c);
return;
}
```

/* Function: write_data - writes an int, float and char */

```
void write_data(struct set set_3)
{
fprintf(stdout,"Supplied data values are: %d %f %s\n",
                        set_3.a, set_3.b, set_3.c);
return;
}
```

Considering *main*, having declared *set_1* and *set_1_ptr*, function prototype statements specify that *main* will use the functions *read_data* and *write_data*. The first function prototype statement specifies that a variable of type *struct set* will be passed by reference to *read_data*. The second function prototype statement, for *write_data*, specifies that *main* will pass a variable of type *struct set* by value. Both function prototypes indicate that neither function will pass data back to *main* via their return statements.

After the address of *set_1* is stored in *set_1_ptr*, *main* calls *read_data* passing *set_1_ptr* to it. Passing the value of *set_1_ptr* means that *set_1* is passed by reference from *main* to *read_data*. When this happens, *read_data* creates a pointer, called *set_2_ptr* and copies the value of *set_1_ptr* into it. When *read_data* calls *fscanf* it uses *set_2_ptr* to access the member variables in *set_1*. Note, however, that the *&* operator is still required to get the address of each member variable. The result of this is that *fscanf* stores the values supplied via the keyboard in the members of *set_1*.

When *read_data* has completed its task, control returns to *main*, which then calls *write_data*, passing *set_1* to it by value. When this happens, *write_data* creates a variable called *set_3*, of type *struct set*, and copies the contents of *set_1* into it. The *write_data* function then passes the members of *set_3* by value to *fprintf* to display them on the screen.

Tutorial 3.8
Implement Program 3.8 and make brief notes on its operation.

3.6 A program to calculate the area of a triangle

Program 3.9 overleaf uses the approach developed in the last section to support the operation of a third function that performs a calculation. The example application considered here is a development of the area calculation program found in Chapter 2, Question 10 in the typical examination questions at the end of the book. In this problem, a triangle is defined by three pairs of *x,y* co-ordinates, supplied via the keyboard.

In addition to *main*, Program 3.9 contains the following functions: *read_points* to get the co-ordinates, *calculate_area* to calculate the area of the triangle and *write_area* to display the area value on the screen. Consider that each of these functions has a specific responsibility within the program, and that *main* is responsible for co-ordinating them.

A structure, *struct triangle*, is defined outside of each function, allowing the *struct triangle* data type to be used in each. It is intended that a variable of this type, *triangle_1*, will store all of the data relating to any triangle specified by the user. The triangle structure contains two arrays of type *double*, one for the *x* co-ordinates and one for the *y* co-ordinates of the vertices. The triangle structure also contains a variable called *area*, of type *double*, in which the calculated area of the triangle is stored.

/* Program 3.9 - Calculating the area of a triangle */

```
#include <stdio.h>
#include <math.h>

struct triangle
  {
  double x[3];
  double y[3];
  double area;
  };

int main(void)
{
struct triangle triangle_1;
struct triangle *triangle_1_ptr;

void read_points(struct triangle *);
void calculate_area(struct triangle *);
void write_area(struct triangle);

triangle_1_ptr = &triangle_1;

read_points(triangle_1_ptr);
calculate_area(triangle_1_ptr);
write_area(triangle_1);

return(0);
}
```

/* Function: read_points - reads x,y values for three points. */

```
void read_points(struct triangle *triangle_2_ptr)
{
fprintf(stdout,"Co-ordinates (x,y) of first point (m.):");
fscanf(stdin,"%lf %lf",     &triangle_2_ptr->x[0],
                           &triangle_2_ptr->y[0]);
fprintf(stdout,"Co-ordinates (x,y) of second point (m.):");
fscanf(stdin,"%lf %lf",     &triangle_2_ptr->x[1],
                           &triangle_2_ptr->y[1]);
```

```
fprintf(stdout," Co-ordinates (x,y) of third point (m.):");
fscanf(stdin,"%lf %lf",      &triangle_2_ptr->x[2],
                             &triangle_2_ptr->y[2]);
return;
}
```

/* Function: calculate_area */

```
void calculate_area(struct triangle *triangle_ptr)
{
double   a,      /* distance between points 1 and 2        */
         b,      /* distance between points 2 and 3        */
         c,      /* distance between points 3 and 1        */
         s;      /* half perimeter of triangle             */

a = sqrt((triangle_ptr->x[1] - triangle_ptr->x[0]) *
         (triangle_ptr->x[1] - triangle_ptr->x[0]) +
         (triangle_ptr->y[1] - triangle_ptr->y[0]) *
         (triangle_ptr->y[1] - triangle_ptr->y[0]));

b = sqrt((triangle_ptr->x[2] - triangle_ptr->x[1]) *
         (triangle_ptr->x[2] - triangle_ptr->x[1]) +
         (triangle_ptr->y[2] - triangle_ptr->y[1]) *
         (triangle_ptr->y[2] - triangle_ptr->y[1]));

c = sqrt((triangle_ptr->x[0] - triangle_ptr->x[2]) *
         (triangle_ptr->x[0] - triangle_ptr->x[2]) +
         (triangle_ptr->y[0] - triangle_ptr->y[2]) *
         (triangle_ptr->y[0] - triangle_ptr->y[2]));
s = (a+b+c)/2.0;
triangle_ptr->area = sqrt(s*(s-a)*(s-b)*(s-c));
return;
}
```

/* Function: write_area */

```
void write_area(struct triangle triangle_3)
{
fprintf(stdout," Calculated area is %lf (m2)\n", triangle_3.area);
return;
}
```

Looking at *main*, a variable, *triangle_1*, and a pointer, *triangle_1_ptr*, are both declared to be of data type *struct triangle*. Following this, there is a function prototype statement for each function that *main* will use. These statements specify that *main* will pass a pointer of data type *struct triangle* to *read_points* and *calculate_area*. The function prototype for *write_area* specifies that *main* will pass a variable, of type *struct triangle*, by value. All of the function prototype statements specify that no data will be returned from any function via their *return* statements.

The first executable statement in *main* copies the address of *triangle_1* to *triangle_1_ptr*. This is followed by statements that call each function in the required sequence. When *read_points* is called, *main* passes the value of *triangle_1_ptr* to it. This passes *triangle_1* by reference to read_points. Note, at this stage, that no data values have been stored in the members of *triangle_1*. However, when *read_points* has finished, the *x* and *y* arrays inside *triangle_1* contain the co-ordinates of three points defining a triangle.

In the next statement, *main* calls *calculate_area*, again passing *triangle_1* by reference. The *calculate_area* function uses the values in the *x* and *y* arrays to calculate the area of the triangle, storing this in the *area* member of *triangle_1*.

Finally, *main* calls *write_area*, this time passing *triangle_1* by value.

Looking at the *read_points* function, when it is called it copies the pointer that is passed to it into *triangle_2_ptr*. This is then used to access the members of *triangle_1* and, using the *&* operator, pass their addresses to *fscanf*. Note how individual elements of the *x* and *y* arrays are accessed.

Looking at the *calculate_area* function, the pointer passed to it is copied into *triangle_ptr*, which is then used to access the values in the elements of the *x* and *y* arrays inside *triangle_1*. This data is used to calculate the length of each side of the specified triangle, the function storing the lengths in local variables *a*, *b* and *c*. These are then used to calculate the half perimeter of the triangle, *s*, before calculating the triangle area. The area value is then stored in the *area* member of *triangle_1*.

The *main* function passes the variable *triangle_1* to *write_area* by value. This means that the values of the members inside *triangle_1* are copied into the members of *triangle_3*. Within *write_area*, the value stored in the *area* member of *triangle_3* is then passed by value to *fprintf*, which displays it on the screen.

Tutorial 3.9
Implement Program 3.9, making brief notes on its operation.

Tutorial 3.10
Re-write Program 3.9 so that the *struct triangle* structure is replaced by

```
struct triangle
{
double point_1[2];
double point_2[2];
double point_3[2];
double area;
};
```

where *point_1[0]* and *point_1[1]*, etc. hold the x and y co-ordinates, respectively, for each point.

Chapter review

This chapter has introduced the detailed 'mechanics' of using functions in C programs. All C programs contain a function called *main*. Many C programs use *main* to co-ordinate the use of other functions, each of these having been designed to perform a particular task. When a function uses another function, the former is said to be the calling function and the latter is the called function. A calling function must have a function prototype statement for each function that it will use. Function prototype statements define the interface between the calling and called functions.

The calling function can give data to the called function through an argument list, variables being passed by value or by reference. In either case, the called function makes a copy of each variable passed to it. A function receiving a variable that has been passed by reference can change the value of that variable in the calling function because both calling and called functions have access to the location in memory of the variable. This cannot happen if the variable has been passed by value.

A called function can pass a variable back to the calling function through the *return* statement. In addition to demonstrating the mechanics of using functions, later examples in this chapter also introduced the use of functions as a means of partitioning a program to reflect the intended solution to a problem.

4

Decisions and Loops

4.1 Introduction

Programs are much more useful if they can make decisions about what tasks need to be performed. Making a decision in a C program usually involves testing the value of one or more variables, for example, 'if X is greater than Y then carry out task 1, else carry out task 2'. C programs can use a range of tests to satisfy many different circumstances. The example given above uses the *if-else* construct and is just about the most simple test that a C program can perform. However, it is not too difficult to imagine that, having made this decision, task 1 or task 2 may also be an *if-else* type of test, leading to the execution of other, more specialized, tasks (perhaps including more tests). This can be achieved in C by using as many **nested** *if-else* constructs as required. The *switch* construct is similar to the nested *if-else* but is more appropriate when different tasks must be selected (switch to) depending on the value of a variable. Another type of test is required when a particular task has to be performed some number of times. If the number of times required is known beforehand then the *for* loop can be used. The decision that has to be made in the *for* loop involves testing a counter to see if the loop has been performed the required number of times. There are other situations where a loop is required, but the number of times that the task has to be carried out is not known beforehand. A simple example of this concerns reading input from the user, where the user can enter as many items of data as they wish. For this and many other similar situations, C provides the *while* and the *do-while* loops.

4.2 The *if-else* construct

The *if-else* construct in C has the following form:

```
if (expression)
    statement₁
else
    statement₂
```

In this example, *(expression)* is a decision that uses relational and logical operators (Section 2.3) to compare the values of variables and constants.

If *expression* is TRUE then its resultant numerical value is 1 and *statement₁* is executed, otherwise *expression* is FALSE, with a numerical value of 0, and *statement₂* is executed. Consider the following:

```
double temperature, set_point = 21.0;

temperature = 32.0;
if (temperature > set_point)
    fprintf(stdout," It's hot today !\n");
else
    fprintf(stdout," It's cold today !\n");
```

In this example, the *if* statement makes a decision by testing for *temperature* greater than *set_point*. If this is TRUE the 'hot' message is displayed. Conversely, the test is FALSE if *temperature* is less than or equal to *set_point*, leading to the 'cold' message being displayed.

Comparison expressions only compare individual numerical values. This means that collections of values (arrays, data structures and character strings) cannot be compared numerically. For example, given:

```
int a[2] = {3, 265}, b[2] = {302, 0};
if (a <= b) .....               is not allowed, but
if (a[1] <= b[1]) ....          is OK because individual elements of
                                each array are compared
```

Similarly:

```
char a[ 3] = "cd", b[3 ] = "tg";
if (a <= b) .....               is not allowed, but
```

if (a[1] <= b[1]) ....   is OK, since ASCII values of individual characters are
                                            compared

Finally, it should be noted that, if a particular decision does not require the *else* part of the *if-else* construct, it need not be included, for example:

int A = 10, B = 9;

if (A < 15)   changes the value of *B* if *A* satisfies the test
  B = 0;

> Tutorial 4.1
> Implement the 'temperature' example as a working program that reads temperature values from the user and displays the appropriate message on the screen. Make notes on its operation.

4.3 Compound statements

In many cases, the outcome of a decision is that a group of statements must be executed. This is achieved by grouping the required statements within '{ ... }' brackets to form a **compound statement** or **block**, for example:

```
int get_data_from_user(struct data *);
int copy_users_data_to_file(struct data *);
int get_data_from_file(struct data *);
struct data *data_ptr;
int user_data;

if (user_data != 0)
  {
  get_data_from_user(data_ptr);
  copy_users_data_to_file(data_ptr);
  }
else
  get_data_from_file(data_ptr);
```

In this example, *user_data* is a **flag** which is used to indicate that data must be read either from the user (if *user_data* is not equal to zero) or from a file (if *user_data* is equal to zero). In this example, if the test is TRUE, data read from the user is stored in a file. This requires the use of two functions, so their calling statements are enclosed within brackets.

In those situations where the comparison expression compares one variable with the value zero, as above, a short hand notation can be used where *if (user_data != 0)* and *if (user_data)* are equivalent. In other words, both tests say 'if *user_data* is not equal to zero then ...'. In contrast to this, the *if (user_data == 0)* test can also be written as *if (!user_data)*, meaning that the comparison expression is TRUE if the value of *user_data* is equal to zero.

Tutorial 4.2
Implement the example in this section as a working program, replacing the function calls with calls to *fprintf*, so that messages indicating the appropriate action are displayed on the screen. The user should be prompted for and supply a value for *user_data* prior to the *if-else* construct.

Tutorial 4.3
Modify the program in the previous question so that it calls the functions shown in the example. Move the calls to *fprintf* into the relevant functions. Note that this involves some understanding of functions, discussed in Chapter 3 and that you will need to invent a convenient definition for *struct data*.

4.4 Nested *if-else* statements

Nested *if-else* statements can be used where multiple decisions must be made. For example, consider the situation where a process controller may perform either of three actions depending on the temperature being above an upper set point, below a lower set point, or in between the two. This can be programmed as follows:

```
if (temperature > upper_set_point)
  {
  report_process("too_hot");
  activate_cooler();
  }
else
  {
  if (temperature < lower_set_point)
    {
    report_process("too_cold");
    activate_heater();
    }
  else
    report_process("OK");
  }
```

In this example, if the outcome of the first test is FALSE then a further test must be performed to decide whether *temperature* is above or below *lower_set_point*. Hence, the statement associated with the *else* part of the first test is another *if* statement. If the outcome of this test is FALSE then the final *else* statement is executed. If the outcome of the first or second test is TRUE then the associated blocks of statements are carried out. When this happens, all of the subsequent statements in the nested *if-else* construct are ignored. Note that the second *if-else* construct is written as a compound statement associated with *else* in the first test. Whilst this is not strictly necessary in this example, nesting *if-else* constructs within compound statements can be important in more complex decision-making, especially when *if-else* constructs and *if* statements are used together.

Computing efficiency can be a significant point when *if-else*, and especially when nested *if-else*, constructs are used within loops. If it is known that a particular outcome is more likely than any other, it is more efficient to place it with the first '*if*'.

Tutorial 4.4
Implement the 'temperature' example in this section as a working program, replacing the function calls with calls to *fprintf*, so that messages indicating the appropriate action are displayed on the screen. The user should be prompted for and supply values for *temperature*, *upper_set_point* and *lower_set_point* prior to the *if-else* construct. Make notes on the program's operation.

Tutorial 4.5
The program in the previous question will not be efficient if the controlled process is predominantly within the upper and lower set points. Assuming this to be the case, re-write the program so that it is more efficient. Hint: you will need to use the logical AND operator, *&&*, discussed in Chapter 2.

Tutorial 4.6
Implement a weather forecasting program that writes the text shown in the rows and columns of the following table, depending on values of temperature and pressure supplied by the user. Define high temperatures to be above 20.0°C and high pressures to be above 1000.0 millibar.

	High pressure	Low pressure
High temperature	Clear skies and hot	Warm with rain
Low temperature	Clear skies and cold	Snow!

4.5 The *switch* construct

The *switch* construct is similar to nested *if-else* statements. However, it is more appropriate than nested *if-else* where different actions are required depending on an integer variable or expression having a range of values. The *switch* construct is defined as follows:

```
switch(expression)
  {
  case constant expression:
                      statements
  case constant expression:
                      statements

    .

    .

  default:
        statements
  }
```

Typically, *expression* in the *switch* statement is the name of an integer variable. Also, the subsequent *case* statements are always enclosed within '{...}' brackets. Each *case* statement consists of the actual word '*case*', followed by an integer constant and a colon. When the *switch* construct is executed, the value of the *expression* in the *switch* statement is compared to the value of the *constant expression* in the first *case* statement. If the two values are equal then the statements associated with the first *case* statement are executed. By default, having executed these statements, processing continues by comparing the *switch expression* with the *constant expression* in each of the subsequent *case* statements. If only those statements associated with a particular *case* statement are to be executed, then the final statement before the next *case* must be *break;*.

The *default* case statement is intended to capture all of the situations that are not trapped by any of the *case* statements. The *default* case is optional and, if omitted, the *switch* construct will not perform any action if none of the *case* constant expressions agree with the expression in the *switch* statement. The following example demonstrates the *switch* construct.

```
int user_reply;
int function_0(void);
int function_1(void);
int function_2(void);
int report_input_error(int);

fprintf(stdout,"Select required function (0,1 or 2):");
fscanf(stdin,"%d", &user_reply);
```

```
switch(user_reply)
  {
  case 0: function_0();
        break;
  case 1: function_1();
        break;
  case 2: function_2();
        break;
  default: report_input_error(user_reply);
  }
```

The code in this example is intended to call particular functions depending on the input supplied by the user and, also, to detect user input that does not correspond to any of the available options. An *int* variable is declared which will be used to store the user's input. Four functions are also declared. The first three of these are to be used depending on the user's choice. Each returns an *int* type value and, for simplicity in this example, has a void argument list. The final function that is declared will be called if the user's input does not correspond to any of the valid options. The first two executable statements prompt the user for their input and store it in *user_reply. user_reply* then appears in the *expression* part of the *switch* statement. This means that the value of *user_reply* will be compared with the values appearing in the *case* statements, starting at the top of the list. Assuming that a match is found, the relevant function will be called. When a function has completed its task, program control will return to the *break* statement immediately following the function call. This will then take the program to the first statement after the *switch* construct. If a match between *user_reply* and any of the *case* statements cannot be found, the default case is activated, calling the *report_input_error* function.

Tutorial 4.7
Implement the above example as a working program. Each of the three functions, *function_0*, *function_1* and *function_2* should simply call *fprintf* to display a message on the screen indicating which option has been selected. The *report_input_error* function should display a suitable error message.

Tutorial 4.8
Modify the program in Tutorial 4.7 so that *user_reply* is passed to each function and its value is displayed in the messages generated via *fprintf*.

4.6 The *for* loop

The *for* loop has the general form:

```
for(initialization expression;
    termination expression;
    incrementation expression) statement
```

The intention of the *for* loop is to execute *statement* a fixed number of times. When the program arrives at the loop for the first time, the decision making within the loop is carried out in three steps:

- initializing a counter in the *initialization expression*;
- using the *termination expression* to check that the counter has or has not exceeded a limiting value;
- after executing *statement*, incrementing the counter using the *incrementation expression*.

The *termination expression* and the *incrementation expression* then control second and subsequent passes around the loop. As an example, Program 4.1 sums the elements of an array.

```
/* Program 4.1 - Array summation */

#include <stdio.h>

int main(void)
{
int a[10] = {1,2,3,4,5,6,7,8,9,10};
int no_numbers = 10, sum = 0, i;

for(i=0; i < no_numbers; i++)
    sum += a[i];
```

```
fprintf(stdout,"Sum of array elements = %d\n", sum);

return(0);
}
```

Program 4.1 declares and initializes an integer array, a, containing 10 elements. Three other variables are also declared. Firstly, *no_numbers* specifies the number of elements in a that contain useful data. Secondly, *sum* will be used as an accumulator and is initialized to zero. Thirdly, the variable i will be used as the counter in the *for* loop. The first executable statement is the *for* construct. Its use in Program 4.1 should be compared to its general form, above. In Program 4.1, the *initialization expression* sets the loop counter, i, to zero. The counter is then compared with *no_numbers* in the *termination expression*. Since this comparison is TRUE, the statement *sum += a[i];* is executed. This simply adds together the value of the current (i^{th}) array element and the current value of *sum*, storing the result back into *sum*. Following this, i is incremented by 1 in the *incrementation expression*. This is the end of the first pass around the *for* loop. At the start of the second pass, i is compared to *no_numbers* in the *termination expression*. If this decision is TRUE, the controlled statement is repeated, and so on.

Note that the *termination expression* tests for i less than *no_numbers*. This choice of *initialization* and *termination expressions* is guided by the use of array, a, in the statement that the loop controls. Remember that in C array elements are indexed from 0. So, the 10th element of a will be element 9. Hence, the loop should stop after it has processed the last element, when i is equal to *no_numbers*.

In Program 4.1 only one statement is executed under the control of the *for* loop. In situations where several statements need be executed, they must be enclosed within '{ .... }' brackets to form a compound statement, as shown in Program 4.2.

```
/* Program 4.2 - A fruit identification program */

#include <stdio.h>

int main(void)
{
char fruit_names[20][20];
int max_no_fruit = 20, no_fruit, i;
```

```
fprintf(stdout,"How many fruit can you name (maximum %d):",
        max_no_fruit);
fscanf(stdin,"%d", &no_fruit);

for(i=0; i < no_fruit; i++)
  {
  fprintf(stdout,"Name of fruit:");
  fscanf(stdin,"%s", fruit_names[i]);
  }
fprintf(stdout,"You have named the following fruits:\n");
for(i=0; i < no_fruit; i++)
  fprintf(stdout,"%s\n", fruit_names[i]);

return(0);
}
```

Although Program 4.2 may appear a little silly in asking the user to input the names of fruits, it conveniently demonstrates the use of compound statements in loops and also reveals several important features of the *for* loop. The program uses a *for* loop that controls two statements, a call to *fprintf* to prompt the user, followed by a call to *fscanf* to store the user's input in the array *fruit_names*. Note that the *fruit_names* array is declared to store a maximum of 20 names, each containing up to 19 characters. Before executing the *for* loop, the program needs to know how many times the loop will be executed. This information is obtained from the user via the calls to *fprintf* and *fscanf*, prior to the *for* statement.

Note how the message displayed by the first call to *fprintf* tries to limit the maximum number of names supplied by the user, so that they don't enter more names than *fruit_names* can store. The user can, however, enter any integer value regardless of the displayed message. Suppose that the user types 100. This value is stored in *no_fruit*, which also appears in the termination expression of the *for* loop. This means that the statements controlled by the loop will attempt to prompt the user to supply 100 names, trying to store them in an array having enough space for just 20. Entering the 21st name will corrupt the program. To trap this problem it could be argued that an additional *if-else* construct could be inserted before the *for* loop. However, this misses two very important points. Firstly, if the user can supply more than 20 names, the initial choice of 20 for the size of the *fruit_names* array forces an arbitrary and unnecessary limit on the user. Secondly, if any value greater than 20 is stored in *no_fruit*, information in the *for* loop *termination expression* will not be consistent with the statements that the loop controls, eventually leading to an uncontrolled failure of the program.

Another problem that will occur is that the user must actually count the number of fruit that they think they can name before actually naming them. This is inconvenient for the user because they essentially have to do the same job twice. Also, suppose that the user elects to name 15 fruits, but can then only think of 7. The *for* loop will not end until 15 character strings have been supplied by the user. Thus, for the program to continue, the user must invent some names. From this, it should be clear that, although *for* loops can be used for controlling interaction with the user, they are not generally appropriate.

for loops are most appropriate where all of the information required to control the loop is known completely before the loop executes and the programmer can ensure that there will be no conflicts between the information used to control the loop and the information used by the statement(s) within the loop, as was the case in Program 4.1. Another example of this is shown in Program 4.3, which prints out a table of trigonometric function values.

```
/* Program 4.3 - Trigonometric function display */

#include <stdio.h>
#include <math.h>

int main(void)
{
double x, pi = 22.0/7.0;

fprintf(stdout," x    sin(x)  cos(x)  tan(x)\n");
for(x=0.0; x <= pi/2.0; x+= 0.1)
    fprintf(stdout,"%lf %lf %lf %lf\n", x, sin(x), cos(x), tan(x));

return(0);
}
```

In Program 4.3, a variable, *x*, of type *double* is declared and then used as a counter in the *for* loop. The same declaration statement initializes *pi* using the result of a calculation so that its value is as accurate as the precision of the *double* data type will allow. Prior to the *for* loop, *fprintf* is called so that four column titles will appear on the screen. When the *for* loop starts, *x* is initialized to zero and then compared to *pi/2.0*. Since the comparison is TRUE, *fprintf* is called to display the trigonometric values for the current value of *x*. *x* is then incremented by 0.1 radians and is again compared with *pi/2.0* in the *termination expression*. The loop stops when the value of *x* exceeds the value of *pi/2.0*. Since *sin( )*, *cos( )* and *tan( )* are functions, *main* must contain prototype statements for them. These are contained in the standard *math.h* library which is copied into the program using the *#include <math.h>* statement.

Tutorial 4.9
Implement Program 4.1 and make brief notes on its operation.

Tutorial 4.10
Change the value of *no_numbers* from 10 to 5 in Program 4.1 and explain how and why the output has changed.

Tutorial 4.11
Implement Program 4.2, making brief notes on its operation. Supply various inputs to the program to make it fail in each of the ways discussed.

Tutorial 4.12
Insert an *if-else* construct just before the *for* loop in Program 4.2 that will stop the user from entering more than 20 names. After demonstrating that this works, run the program again and enter –1 when prompted. Diagnose and fix this problem.

Tutorial 4.13
Implement Program 4.3 and make brief notes on its operation, especially concerning how the trigonometric values are evaluated and displayed on the screen.

4.7 The *while* loop

The *while* loop has the general form:

> while (continuation expression)
>   statement

continuation expression is a decision which is first evaluated before the *while* loop is entered. If *continuation expression* is FALSE when the program arrives at the loop, the program jumps over the loop and ignores it. If *continuation expression* is TRUE then the first iteration around the loop is made. *continuation expression* is then tested again before the start of each subsequent iteration. The loop terminates when *continuation expression* is FALSE. If it is never FALSE, the *while* loop never stops. An example of a *while* loop is given in Program 4.4, which is a simple arithmetic game.

```
/* Program 4.4 - The numbers game */

#include <stdio.h>
#include <stdlib.h>

int main(void)
{
int number = 10000, difference = 0;

fprintf(stdout,"The NUMBERS game\n");
fprintf(stdout,"———————————\n");
fprintf(stdout,"Subtract the numbers shown from 10000\n");
while(number+difference == 10000)
  {
  number = rand();
  fprintf(stdout,"Number: %d    Difference:",number);
  fscanf(stdin,"%d", &difference);
  }
fprintf(stdout,"You either got it wrong or you wanted to stop !\n");

return(0);
}
```

The objective of the game in Program 4.4 is that the program gives the user a random integer which the user must subtract from 10,000. To stop the program the user simply supplies a wrong answer. Having declared *number* and *difference* as integers, the program then provides some instructions for the user. For the *while* loop to start, its *condition expression* must be TRUE. This has been ensured by the choice of initial values for *number* and *difference*. Inside the *while* loop, C's random number generator function, *rand*, is used, storing the returned value in *number*. This is then displayed on the screen. The user's reply is read in the final executable statement inside the loop. The *continuation expression* in the *while* statement compares the sum of the random and user-supplied numbers with the target value of 10,000. If the sum is equal to the target value, the *continuation expression* is TRUE and the loop is executed again. In passing, observe that the random number generator function, *rand*, can only be used if the *#include <stdlib.h>* statement is present to provide the necessary function prototype statement.

It is important to note that the *while* loop will continue as long as the user is willing to play the game. Hence, the program cannot know, before the loop starts, how many times the loop will be executed. This should be contrasted with the limitations of the *for* loop, discussed in the previous section. Program 4.5, below, high-lights the contrast by re-coding the 'fruit' program, Program 4.2, from the previous section. In Program 4.5 repetitive interaction with the user is controlled by a *while* loop whose *continuation expression* compares the user's reply to a character string constant, '*end*'. This comparison is carried out using a standard C function called *strncmp*, which compares the first three characters stored in *reply* with the string constant '*end*'. Note that, to use *strncmp* and *strcpy*, below, the *string.h* library must be included at the start of the program. In this program, the criterion for the loop to continue is that any of the first three characters in *reply* should be different to those in '*end*'. This must also be the case for the loop to start, so *reply* is initialized to a blank or empty string when it is declared.

/* Program 4.5 - The fruit program again */

```
#include <stdio.h>
#include <string.h>
```

```
int main(void)
{
char fruit_names[20][20];
char reply[20] = "";
int max_no_fruit = 20, no_fruit = 0, i;

while(strncmp(reply,"end",3) != 0)
  {
  fprintf(stdout,"Name of fruit or 'end' ?:");
  fscanf(stdin,"%s", reply);
  if (strncmp(reply,"end",3) != 0)
    strcpy(fruit_names[no_fruit++], reply);
  }
fprintf(stdout,"You have named the following fruit:\n");
for (i=0; i < no_fruit; i++)
  fprintf(stdout,"%s\n", fruit_names[i]);


return(0);
}
```

Two other differences should be apparent between Programs 4.5 and 4.2. The first is that, in Program 4.2, the value of variable *no_fruit* is used to determine how many times to go around the *for* loop. In contrast, in Program 4.5 *no_fruit* is initialized to zero and is incremented by one inside the *while* loop every time that the user replies to the program. Hence, *no_fruit* is now used as a counter rather than a limit. The second difference is the need for a decision inside the *while* loop but not inside the *for* loop. This decision is needed to work out whether the user's latest input is simply more data or an indication that no more data will be supplied. If the user's input is not '*end*' then it is data and must eventually be stored in the *fruit_names* array. If the user has typed '*end*' then their input must not be stored in the array.

This is a significant change from Program 4.2. In Program 4.5 input from the user is stored in *reply*. Since character strings are, themselves, arrays it is not possible to copy *reply* to the *fruit_names* array by simply using the assignment operator. Each character in the string must be assigned individually. C provides a standard function, *strcpy*, to do this. The first argument given to *strcpy* is the next empty element of the *fruit_names* array which is identified by the current value of *no_fruit*. In passing the value of *no_fruit* to *strcpy*, note how *no_fruit* is followed by the incrementation operator which adds one to *no_fruit* after it has been used to index the *fruit_names* array.

From the user's point of view, the major advantage of Program 4.5 over Program 4.2 is that they do not have to specify the number of names before they supply the names themselves. This removes the need for the user to have to 'think ahead' and also removes the possible difficulty of users electing to name perhaps 15 fruits but only being able to name seven. However, Program 4.5 still has one major drawback. Whilst users can enter as many names as they wish, the program can still store only 20 names in *fruit_names*. Entering more than 20 names will still cause the program to fail. The easiest, although not the most general, way to overcome this is by using a *continuation expression* in the *while* statement that makes two decisions. For example, changing

```
while(strncmp(reply,"end",3) != 0)
```

to

```
while((strncmp(reply,"end",3) != 0) && (no_fruit < max_no_fruit))
```

would allow the *while* loop to continue provided that both tests are TRUE and would cause the loop to terminate either if the user typed '*end*' or the user tried to enter more than *max_no_fruit* fruit names.

Tutorial 4.14
Implement Program 4.4 and make notes on its operation.

Tutorial 4.15
Implement Program 4.5 and make notes on its operation.

Tutorial 4.16
Modify the program in the previous question so that it will not allow the user to supply more than four names.

4.8 The *do-while* loop

The *do-while* loop differs from the *while* loop in that the *continuation expression* is evaluated at the end of each iteration around the loop, rather than at the beginning. This means that the statements within a *do-while* loop are always executed at least once. As an example of this, in Program 4.6 below, the *while* loop used in Program 4.5 has been replaced with a *do-while* loop. One further change is that the *reply* character string no longer needs to be initialized when it is declared since, by the time it is used as part of a decision, it has been given a value by the user.

```
/* Program 4.6 - The fruit program yet again */

#include <stdio.h>
#include <string.h>

int main(void)
{
char fruit_names[20][20];
char reply[20];
int max_no_fruit = 20, no_fruit = 0, i;

do
  {
  fprintf(stdout,"Name of fruit or 'end' ?:");
  fscanf(stdin,"%s", reply);
  if (strncmp(reply,"end",3) != 0)
     strcpy(fruit_names[no_fruit++], reply);
  }
  while(strncmp(reply,"end",3) != 0);

fprintf(stdout,"You have named the following fruit:\n");
for (i=0; i < no_fruit; i++)
  fprintf(stdout,"%s\n", fruit_names[i]);

return(0);
}
```

Tutorial 4.17
Implement Program 4.6 and make notes on its operation.

Chapter review

Facilities for decision making are vital elements of any programming language. C provides a range of decision-making constructs that are not only very flexible in their own right, but can also be combined in many ways to meet the needs of any programming task. This functionality does, however, come at the price of the programmer needing a sound knowledge of how each works, and which, possibly in combination, will provide the best solution to a problem. In turn, this relies on the programmer having a good understanding of the problem, the different possible approaches to its solution and the way in which the user is expected to interact with the resulting software.

This chapter has demonstrated that simple choices between alternatives can be achieved through the *if-else* construct, possibly repeated or nested for more complex decisions. In many situations, the *switch* construct provides a more structured and more easily understood alternative to the nested use of *if-else*. It has also been shown that decision making is a key element of a program repeating a task. In using a *for* loop, a counter that is part of the *for* construct keeps count of the number of times that the controlled task has been carried out, deciding to terminate the loop by comparing the counter with a limit in a *termination expression*. In the *while* and *do-while* loops the program must check the continued truth of a *comparison statement* in order to execute another pass around the loop. Unlike the *for* loop, this often involves comparing a counter or some other indicator that is programmed to change its value as an integral part of the controlled task. This difference between the *for* and *while* or *do-while* loops has highlighted several implications for their appropriate use in different circumstances.

5

Files and Formatting

5.1 Introduction

Many engineering programs read the data that they are required
to process from one or more **files** on disc and, having carried out
their task, write their results to other files. To do this, links must be
established between the program and the relevant files. In C these
links are called **streams**. Two examples of streams that you have
already used extensively are *stdin* and *stdout*, for reading input
from the keyboard and writing data to the screen, respectively. The
first part of this chapter concentrates on the basic mechanics of
transferring relatively simple collections of data between
programs and files. Subsequent sections develop more comprehensive approaches to input and output that are needed when
relatively large amounts of complicated data need to be either
read from or written to files. These sections will concentrate on the
use of formatting codes, together with line input and output. You
will also see that decision making and loops also have significant
roles to play.

5.2 Reading and writing

To read from and write to files we need to create streams, like *stdin*
and *stdout*, but having names that we choose. Consider the following
statements:

```
FILE *input;
int A;

input = fopen("input.dat", "r");
if (input != NULL)
  {
  fscanf(input,"%d", &A);.
  fclose(input);
  }
```

Here, the first statement declares a stream, naming it *input*. *FILE* is a data type provided by C to create streams. It may seem strange to use *FILE* when we mean 'stream' but for a program to use a stream, it must also use several other variables that C usually keeps hidden. These variables, along with the stream are all contained inside a data structure of type *FILE*, defined in the *stdio.h* library. So, in the above example, we declare *input* and call it a stream, but in actual fact *input* is a pointer to a data structure of type *FILE*. In the first executable statement a standard C function, *fopen*, is used to make the link between the program and a file called '*input.dat*'.

Notice that *fopen* has to be given two input arguments. The first is a character string that names the file to be linked to. The second is a single character that describes the type of link. In the above example, this character is '*r*' which means that the program is allowed to **read** data from '*input.dat*'. If the *fopen* function successfully makes the link between the program and the file, it creates a data structure of type *FILE* and returns a pointer holding its address. In this example, this address is then copied into *input*. If *fopen* cannot make the link, it returns the *NULL* value, indicating that it has failed. As shown, the success or failure of *fopen* is usually tested in an *if* statement. If the stream has been successfully created, the statements in the { ... } are carried out. The first of these calls *fscanf* to read an integer value from the file and store it in variable *A*; having done this, the link between the program and the file is broken by calling the *fclose* function to close the stream. It should be noted that the function prototype statements needed for *fopen* and *fclose* are provided in the *stdio.h* file, included at the start of the program.

For writing data to a file we could use the following example:

```
FILE *output;
int A = 10;

output = fopen("output.dat", "w");
if (output != NULL)
  {
  fprintf(output, "A = %d\n", A);
  fclose(output);
  }
```

Again, the first declaration statement creates a stream, this time called *output*. The first executable statement uses *fopen* to make a link between the program and the file, using '*w*' to allow the program to **write** to the file. By using '*w*' writing always starts at the beginning of the file. If the file already contains data, this will be lost when the new data is written. If we do not want this to happen, we can use '*a*' instead of '*w*', forcing any new data that the program sends to the file to be **appended** or added to the end of any data that is already there. In the above example, if *fopen* is successful in making the link, *fprintf* is used to write the value of *A* to the file before the stream is closed using *fclose*.

Program 5.1 shows how streams to and from files are used.

```
/* Program 5.1 - Reading from and writing to files            */
/*                                                            */
/* Reads an integer, double and a character string from the keyboard.   */
/* Opens a stream to a file called file1.dat, stores the values in it and   */
/* closes the stream.                                         */
/* Re-opens the stream to the same file, this time for reading, reads   */
/* the values from the file into new variables and displays the values   */
/* on the screen.                                             */

#include <stdio.h>

int main(void)
{
int A, D;
double B, E;
```

```
char C[101], F[101];
FILE *in_stream, *out_stream;

fprintf(stdout, "Enter an Integer, Real and a Character string:");
fscanf(stdin, "%d %lf %s", &A, &B, C);

out_stream = fopen("file1.dat", "w");
if (out_stream != NULL)
  {
  fprintf(out_stream, "%d  %lf  %s\n", A, B, C);
  fclose (out_stream);
  }

in_stream = fopen("file1.dat", "r");
if (in_stream != NULL)
  {
  fscanf(in_stream, "%d %lf %s", &D, &E, F);
  fclose (in_stream);
  }
fprintf(stdout, "Values stored and then retrieved are: %d %lf %s\n",
D, E, F);

return(0);
}
```

Program 5.1 uses four streams, *stdin*, *stdout*, *in_stream* and *out_stream*. In the first executable statement the user is prompted to supply three inputs, an integer, a real and a character string. Having read these, the program then uses *fopen* to create *out_stream*, connecting the program to file *file1.dat*. Note that the 'w' argument opens the file for writing. If the link has been successfully made, *fprintf* is used to write the user's input into the file using *out_stream*. Closing the file using *fclose* then breaks the link between the program and *file1.dat*. Next, *file1.dat* is re-opened using *in_stream*, this time using 'r' to open it for reading. If the file has been opened successfully, *fscanf* is used to read the previously written data into different variables before closing the file using *fclose*. The program then writes the values read from the file to the screen.

Another example is shown in Program 5.2, where file input and output is used in a program that consists of several functions. Program 5.2 is a development of the program in Chapter 2,

Question 10 in the typical examination questions at the end of the book, whose task is to calculate the area of a triangle defined by any three (*x, y*) points. The major difference between Program 5.2 and the previous version is that the former reads the point co-ordinates from a file and writes the calculated area to another file. To do this in a well-structured manner, Program 5.2 contains an additional function to read the names of the files from the user into a data structure. Variables are passed to functions either by reference if the function is going to change their value, or by value if a function simply uses their value.

```
/* Program 5.2 - Calculating the area of a triangle          */
/* Reads the names of the input and output files from the user.   */
/* Reads (x, y) for each point from the input file.          */
/* Calculates the area of the triangle.                 */
/* Writes the area value to the output file.             */

#include <stdio.h>
#include <math.h>

struct triangle
  {
  double x[3];
  double y[3];
  double area;
  };

struct file_names
  {
  char input_filename[101];
  char output_filename[101];
  };

int main(void)
{
struct file_names io, *io_ptr;
struct triangle example, *example_ptr;

int read_filenames(struct file_names *);
int read_points(char [], struct triangle *);
```

```
int calculate_area(struct triangle *);
int write_area(char [], double);
io_ ptr = &io;
example_ ptr = &example;

read_filenames(io_ptr);
read_ points(io.input_filename, example_ptr);
calculate_area(example_ ptr);
write_area(io.output_filename, example.area);
return(0);
}
```

```
/* Function: read_filenames; Used in program 5.2.           */
/* Reads two file names as char strings into a structure     */
/*  passed by reference by the calling  function.            */

int read_filenames(struct file_names *filenames_ ptr)
{
fprintf(stdout, "Enter name of input file:");
fscanf(stdin, "%s",&filenames_ptr->input_filename);
fprintf(stdout, "Enter name of output file:");
fscanf(stdin, "%s", &filenames_ptr->output_filename);
return(0);
}
```

```
/* Function: read_points; Used in program 5.2.              */
/* Uses file name passed by value from calling               */
/* function, along with triangle structure, passed           */
/* by reference, to read and store x & y co-ordinates        */
/* of  three points.                                         */

int read_points(char filename[], struct triangle *triangle_ptr)
{
FILE *input;

input = fopen(filename, "r");
fscanf(input, "%lf %lf", &triangle_ptr->x[0], &triangle_ptr->y[0]);
fscanf(input, "%lf %lf", &triangle_ptr->x[1], &triangle_ptr->y[1]);
```

```
fscanf(input, " %lf %lf", &triangle_ptr->x[2], &triangle_ptr->y[2]);
fclose(input);
return(0);
}
```

```
/* Function: calculate_area; Used in program 5.2.        */
/* Calculates area of a triangle defined by three        */
/* points, coordinates supplied by reference in          */
/* a structure pointed to by triangle_ptr.               */
/* Calculated area passed back in same structure.        */

int calculate_area(struct triangle *triangle_ptr)
{
double   a,        /* distance between points 1 and 2        */
         b,        /* distance between points 2 and 3        */
         c,        /* distance between points 3 and 1        */
         s;        /* perimeter/2                            */

a = sqrt((triangle_ptr->x[1] - triangle_ptr->x[0]) *
         (triangle_ptr->x[1] - triangle_ptr->x[0]) +
         (triangle_ptr->y[1] - triangle_ptr->y[0]) *
         (triangle_ptr->y[1] - triangle_ptr->y[0]));

b = sqrt((triangle_ptr->x[2] - triangle_ptr->x[1]) *
         (triangle_ptr->x[2] - triangle_ptr->x[1]) +
         (triangle_ptr->y[2] - triangle_ptr->y[1]) *
         (triangle_ptr->y[2] - triangle_ptr->y[1]));

c = sqrt((triangle_ptr->x[0] - triangle_ptr->x[2]) *
         (triangle_ptr->x[0] - triangle_ptr->x[2]) +
         (triangle_ptr->y[0] - triangle_ptr->y[2]) *
         (triangle_ptr->y[0] - triangle_ptr->y[2]));
s = (a + b+ c)/2.0;
triangle_ptr->area = sqrt(s*(s-a)*(s-b)*(s-c));
return(0);
}
```

```
/* Function: write_area; Used in program 5.2.       */
/* Writes value of area, passed from calling         */
/* function by value, to file named by filename      */
/* which is also passed by value.                    */

int write_area(char filename[], double area)
{
FILE *output;

output = fopen(filename, "w");
fprintf(output, "Area of triangle = %lf\n", area);
fclose(output);
return(0);
}
```

Program 5.2 defines two structures, *struct triangle* and *struct file_names*. Remember that these statements define these structures to be new data types that can be used in declaration statements. Also, remember that these structures are defined outside of any function, so that they provide a common definition available to all functions within the program. This supports the specification of interfaces between calling and called functions via the function prototype statements in *main*. Each function prototype specifies that the relevant function will receive either the value of a member within one of the data structures or it will receive the address of a structure via an appropriate pointer

The first executable statements in *main* store the addresses of *io* and *example*, in the relevant pointers. The remaining statements then call each of the functions as required. In calling *read_file-names*, *main* passes *io* by reference. This allows *read_filenames* to store the names of the input and output files, supplied by the user, in *io*. When *main* calls *read_points*, it passes the name of the input file by value and *example* by reference. The *read_points* function then uses the file name to open the input file by creating a stream called *input*. Having done this, *read_points* reads the co-ordinates of three points and uses its copy of *example_ptr* (called *triangle_ptr* in *read_points*), to store them in *example*. Before leaving *read_points*, the input file is closed because it is no longer needed. Function *main* then calls *calculate_area*, which has not needed to be modified from the program considered in Chapter 3. Finally, *main* calls *write_area*, passing both the name of the output file and the calculated area by value. Passing by value is used, rather than passing by reference, because there is no intention that *write_area* will change the value of either variable passed to it. The *write_area* function uses the output file name to open the output file, storing the stream in *output*. It then writes the value of *area* to the output file before closing the stream.

It is important to note that Program 5.2 differs from the program considered in Chapter 3 only in the source and destination of the data passing through the program. This has required changes to be made only to those parts of the program that are affected by the different requirements. Both programs have to calculate the area of a triangle, so no changes were required to the *calculate_area* function.

Tutorial 5.1
Implement Program 5.1 and make brief notes on its operation.

Tutorial 5.2
Implement Program 5.2 and make brief notes on its operation.

5.3 Formatted output

The C language provides a range of facilities for controlling the style of output produced by programs. You should now be familiar with statements such as:

```
int I = 7;
float A = 9.5;
fprintf(stdout, "%d %f", I, A);
```

where %d and %f are the default formatting codes needed for variables of type *int* and *float*, respectively. These formatting codes are the minimum required in terms of specifying how data values are to be formatted.

Other frequently used options are available, stated generally as

Integer:	*%md, %mu, etc.*	*See Section 1.3*
float:	*%m.pf*	*Decimal notation*
float:	*%m.pe*	*Exponential notation*
double:	*%m.plf*	*Decimal notation*
double:	*%m.ple*	*Exponential notation*

where m is the minimum field width (minimum number of digits to be displayed) and p is the precision, i.e. the number of digits after the decimal point (default 6).

For example, given:

```
int A = 123;
double B = 7.5847e-6;
```

various combinations of m and p give the following results:

fprintf(stdout, "A = %d", A);	*gives A = 123*
fprintf(stdout, "A = %6d", A);	*gives A = 123*
	(3 spaces before the number)

fprintf(stdout, "B = %le", B);	*gives B = 7.458700e-06*
fprintf(stdout, "B = %11.2le", B);	*gives B = 7.46e-06*
	(3 spaces before the number)

The formatting codes discussed here are, perhaps, the most generally useful ones that are needed with *fprintf*. There are others that control all remaining aspects of *fprintf* functionality. Details of these can be found via the Help command in your programming environment.

Tutorial 5.3
Change Program 4.3 in Chapter 4 so that the displayed values are accurate to two decimal places.

Tutorial 5.4
Given the following values, write them to a file on a single line. Each written value should be accurate to three decimal places with two spaces between the first two values, three spaces between the second and third, and so on.

1.54549873e4, 9.28684e3, 687.3577, 3.775559

Having written the values, close the file. Re-open the file for reading and read the values back into the program using different variables. Display, on the screen, each original value, the corresponding value read back from the file and the difference between them. Use the default formatting codes to display all values on the screen. Explain the displayed differences.

5.4 Line output

There is another function, called *sprintf*, which is closely allied to *fprintf*, in that *sprintf* is used to write data to a character string rather than using a stream to write to the screen or a file. There are two primary reasons that we may want to do this. Firstly, if our software is intended to generate lots of output, possibly in different arrangements, it may be more convenient and quicker to assemble the items to be outputted on each line into a character string before writing each completed line to the screen or a file using *fprintf*. The reason for this is that communication between the processor and disc, screen, keyboard, printer, etc., is very slow in comparison to the speed at which the processor and memory operate. Using *sprintf* to put values into a character string is a lot faster than using *fprintf* to write each value to the screen or a file.

The following example shows how *sprintf* can be used to assemble the values of three variables into a single string that is then output to a file. Note how the use of a character string, *line*, with *sprintf* parallels the use of a stream, such as *output*, with *fprintf*.

```
int A = 1, B = 2, C = 3;
FILE *output;
char line[101];
```

```
output = fopen("filename.dat", "w");
sprintf(line, "A = %d B = %d C = %d", A, B, C);
fprintf(output, "%s\n", line);
fclose(output);
```

The second major reason for sometimes using *sprintf* is to generate character strings that are needed within a program. This often concerns the dynamic creation of filenames. Note, in the above example, that a file name, *'filename.dat'*, has been specified as a literal string of characters. Sometimes, especially in iterative calculations, we may want to be more flexible than this and read from or write to collections of files, whose names form a pattern, e.g. 'T1.dat', 'T2.dat', 'T3.dat' ... 'Tn.dat'. Rather than store a list of these names, they can be generated when required by using *sprintf*. For example:

```
int A = 1, B = 2, C = 3;
int I = 0;
char text[3] = "out";
char name[101], line[101];
FILE *output;
sprintf(name, "%s%d.dat", text, I);
output = fopen(name, "w");
sprintf(line, "A = %d B = %d C = %d", A, B, C);
fprintf(output, "%s\n", line);
fclose(output);
```

In the above statements, *sprintf* is used to assemble the name of a file, *'out0.dat'*, from three components – the characters *'out'*, stored in a character string, a numerical value stored in an integer and a group of literal characters *'.dat'*. The assembled file name is then passed to *fopen* by supplying *name* as its first argument.

Tutorial 5.5
Modify the program written for Tutorial 5.4 so that it writes the values to a character string prior to writing the character string to the file.

Tutorial 5.6
Guided by the previous example of using character strings to assemble file names, write a program that uses a *for* loop to get three words from the user, storing each word in a separate file. Use the value of the counter in the *for* loop as part of each file name to ensure that three unique files are created.

5.5 Line input

This section is primarily concerned with three functions. Firstly, *fscanf* and some additional useful features of it. Secondly, *fgets* for reading a whole line from a file and, thirdly, *sscanf* for reading values from a character string. Statements involving *fscanf*, such as the one below should now be familiar:

```
fscanf(input, "%d %lf", &A, &B);
```

where an integer value is read into *A* and a decimal is read into *B*.

If only the second value in this example were required, i.e. we wanted to read and discard the integer value, we could either use the above call to *fscanf* and then ignore *A* or use a modified call to *fscanf* as shown here:

```
fscanf(input, "%*d %lf", &B);
```

Note that the first formatting code is now %*d. This causes *fscanf* to read the integer value but not store it in a variable. Since only the second value will be stored, only the address of *B* has been given to *fscanf*. Similarly, if we wanted to read and ignore a character string, we would use %*s and if we wanted to read and ignore a value of type *double* we would use %*lf. This ability to read and ignore selected values can be very useful when we only want to input particular parts of files. For example, if a file contains a table with column headings, we could read and ignore each heading and then continue to read the values listed in the table.

A feature that *fscanf* shares with *fprintf* is that it returns a value to the calling function. This was not mentioned earlier for *fprintf* because it is not a generally useful facility. However, it can be very useful when using *fscanf*. For *fscanf*, the returned value specifies the number of data items that have been successfully read. In the example below, assuming that an integer and a decimal value are available to be read, *fprintf* would display the value 2:

```
int number_of_inputs;
number_of_inputs = fscanf(input, "%d %lf", &A, &B);
fprintf(stdout, "%d\n", number_of_inputs);
```

This can be a very useful mechanism for detecting input errors, or conditions that may not be errors but the program needs to know about, such as detecting the start of a particular section of input or reaching the end of a file. However, it is important to recognize that *fscanf* needs to be told precisely what to read, which may sometimes be a problem. For example, imagine that an input file can legitimately contain one, two or several tables of data. When a program is needed to read this data, it is better to design it so that it reads whatever combination of tables are present, rather than read all possible tables in a fixed sequence. To do this, the program must be able to read something that identifies that a particular table is present and then do what is required to read that table. The *fscanf* function cannot easily provide that degree of flexibility. For this type of situation (which is actually quite common in engineering and scientific problems) C provides the *fgets* and *sscanf* functions. The *fgets* function is used to read a whole line from a file (defined as everything from the current position in a file up to the next end-of-line character, '\n') into a character string. The *sscanf* function does the reverse of *sprintf*, by reading values from a character string, such as that produced by *fgets*. The following C statements show the essential features of *fgets*.

```
char line[101];
char *line_ptr;
FILE *input;
input = fopen("filename.dat", "r");
line_ptr = fgets(line, sizeof(line), input);
fclose(input);
```

The statements in the above example open a file, called *filename.dat*, and use *fgets* to read the first line from it before closing the file. The *fgets* function reads whatever is contained on the line and stores it in a character string, here called *line*. To do this, *line* must be declared with enough space to hold the longest line that is anticipated. In the above example, *line* is big enough to hold 100 characters (remember that we need an extra character for C to insert the 'end of string' terminator). The *fgets* function needs to be given the character string where it will put the input data, along with the amount of space that is available in that string (most easily provided using *sizeof*), and the name of the stream that connects the program to the file containing the required data.

The *fgets* function returns a pointer, the value of which indicates that *fgets* was successful or not. If the call to *fgets* is successful, the returned pointer holds the address of the *line* character string that is supplied to *fgets*. If the call to *fgets* is not successful, maybe because the file is empty or the end of the file has been reached, the returned pointer has the value *NULL*.

Having used *fgets* to read a line of data into a character string, such as *line*, in the previous example, the *sscanf* function can then be used to read individual data items from *line*. For example, if the file in the above example contained

```
1.1, abc, 7
```

the following statements, employing *fgets* and *sscanf* could be used to read the data:

```
char line[101];
char *line_ptr;
FILE *input;
int A;
char text[4];
double B;
input = fopen("filename.dat", "r");
line_ptr = fgets(line, sizeof(line), input);
fclose(input);
sscanf(line, "%lf %s %d", &B, text, &A);
```

In this example, having used *fgets* to read data into *line*, *line* is then passed to *sscanf* which reads the individual values. Note how *sscanf* uses the name of the character string in the same place that *fscanf* uses the name of a stream.

This approach to reading data from files has two major advantages over using *fscanf*, both resulting from a clearer separation of the 'mechanics' of reading from the interpretation of what has been read. The first benefit is that it becomes easy to search for 'landmarks' in the data such as the start or end of tables and to skip over lines of data that are not required. The second is that detecting the end of a file is easy since *fgets* returns a *NULL* pointer.

The following program aims to show how these benefits are obtained. You will find this a rather ambitious example because it necessarily brings together functions, decision making, loops and input from files. It is reasonably typical of 'industry-scale' software.

Assume that we have a file, called '*fe_mesh.dat*', containing the following:

THIS FILE CONTAINS A NODES TABLE AND AN ELEMENTS TABLE

NODES

Node_id	x	y
1	0.0	0.0
2	0.0	1.0
3	1.0	0.0
4	1.0	1.0
5	2.0	0.0
6	2.0	1.0

THESE TABLES COULD BE VERY LONG AND THERE COULD BE OTHER DATA BETWEEN THEM

ELEMENTS

Element_id	Node_1	Node_2	Node_3	Node_4
1	1	3	4	2
2	3	5	6	4

END OF FILE

Also, suppose that only the numerical parts of the *NODES* table are to be read. Program 5.3 is intended to do this. However, before looking at the program, there are a few points that should be noted about the above data. The *NODES* table does not start at the beginning of the file and, although the *NODES* table contains six lines of data, there is nothing before the table to indicate this. Finally, the end of the data that should be read does not correspond to the end of the file.

Considering Program 5.3, a structure, *struct fe_mesh*, is defined outside of any function, so that all functions can use it as a data type. The *struct fe_mesh* structure has an *int* type member, *no_nodes*, that will be used to count the number of nodes as they are read from the file. For convenience here, the structure also contains various arrays, each large enough to store relevant information for a maximum of 10 nodes. Remember that using arrays is a significant limitation in this and other programs because the size of any array must be arbitrarily specified before compiling. *main* uses the *struct fe_mesh* data type to declare a structure, called *mesh*, and a pointer, *mesh_ptr*. *main* uses the latter to pass *mesh* by reference to the *get_nodes* function, which reads the required data from the file.

```
/* Program 5.3 - Reading and writing a nodes table for a finite element    */
/* mesh                                                                      */

#include <stdio.h>
#include <string.h>
#include <stdlib.h>

struct fe_mesh
  {
  int no_nodes;
  int node_ids[10];
  double x[10];
  double y[10];
  };

int main(void)
{
int return_code = 0;
struct fe_mesh mesh;
struct fe_mesh *mesh_ptr;
int get_nodes(struct fe_mesh *);
int display_nodes(struct fe_mesh);

mesh_ptr = &mesh;
return_code = get_nodes(mesh_ptr);
if (return_code == 0)
  display_nodes(mesh);
else
  fprintf(stdout,"Program unable to read NODES table\n");
return(0);
}
```

/* Function: get_nodes; Used in program 5.3. */

```
int get_nodes(struct fe_mesh *fe_mesh_ptr)
{
int return_code = 0;
int i, no_items_counted;
FILE *input;
char line[101], *line_ptr;

/* open input file */
if ((input = fopen("fe_mesh.dat", "r")) != NULL)
  {
  /* find start of NODES table */
  while (((line_ptr = fgets(line, sizeof(line), input)) != NULL) &&
        (strncmp(line, "NODES", 5) != 0));
  if (line_ptr != NULL)
    {
    fgets(line, sizeof(line), input);

    /* read NODES data */
    i = 0;
    while (((line_ptr = fgets(line, sizeof(line), input)) != NULL) &&
          ((no_items_counted = sscanf(line, "%d %lf %lf",
                                        &fe_mesh_ptr->node_ids[i],
                                        &fe_mesh_ptr->x[i],
                                        &fe_mesh_ptr->y[i])) == 3)) i++;
    if (line_ptr == NULL)
      {
      fprintf(stdout, "End of file while reading NODES table.\n");
      return_code = 3;
      }
    fe_mesh_ptr->no_nodes = i;
    }
  else
    {
    fprintf(stdout, "Error finding start of NODES table.\n");
    return_code = 2;
    }
  fclose(input);
  }
else
```

```
{
fprintf(stdout, "Error opening fe_mesh.dat\n");
return_code = 1;
}
return(return_code);
}
```

/* Function: display_nodes; Used in program 5.3. */

```
int display_nodes(struct fe_mesh mesh)
{
int return_code = 0;
int i;

fprintf(stdout, "NODES table:\n");
for (i=0; i < mesh.no_nodes; i++)
fprintf(stdout, "Node %d  x = %le  y = %le\n",
                    mesh.node_ids[i], mesh.x[i], mesh.y[i]);
fprintf(stdout, "\n");
return(return_code);
}
```

Looking at the *get_nodes* function, *fe_mesh_ptr* is declared in the argument list to hold the copy of *mesh_ptr* that is passed from *main*. The first two declaration statements specify *return_code*, *i* and *no_items_counted* as integers. *return_code* is used to provide feedback to *main*, via the *return* statement, indicating success or failure of the reading process. Variables *no_items_counted* and *i* are used, respectively, to count the number of values read from each line of the file and to count the number of nodes that have been stored. The next declaration statement creates a stream, called *input*, that will connect the program to the *fe_mesh.dat* file. The final declaration statement creates a character string, *line*, and a pointer, *line_ptr*, both of type *char*, which will be used with the *fgets* function.

The first executable statement combines opening the file with deciding what to do if the file cannot be opened. This is achieved by incorporating the call to *fopen* as an operand in the *expression* part of an *if-else* construct (Section 4.2). If *fopen* can open the file the decision is TRUE, allowing the associated block of statements to be executed. If *fopen* cannot open the file, the decision is FALSE. This invokes the corresponding *else*, which displays an error message and sets *return_code* to 1. Assuming that the file is opened successfully, *get_nodes* then uses a *while* loop to search for the start of the *NODES* table. This is done by combining two relational operators and a logical operator within the continuation expression that controls the loop. The first relational operation tests for *fgets* returning a *NULL* pointer.

Remember that *fgets* attempts to read a whole line from a file. If *fgets* is unsuccessful, it returns a pointer having the *NULL* value, in which case the relational operation is FALSE and the *while* loop terminates because, to continue, both relational operations have to be TRUE. If *fgets* is successful, it stores the line that it has read in *line* and returns the address of *line*, which is then stored in *line_ptr*. In this case the first relational operation is TRUE and the second relational operation is then carried out. This operation uses the *strncmp* function to compare the first five characters now stored in *line* with the literal string *NODES*, which is the group of characters appearing at the start of the *NODES* table in the file. The *strncmp* function counts the number of differences between the two character strings. If there are differences, then the line that has just been read by *fgets* is not the start of the *NODES* table. Hence, the second relational operation is FALSE and the *while* loop continues. This means that *fgets* is again used to read the next line.

The *while* loop can stop for two reasons. Either *fgets* reaches the end of the file, or *strncmp* identifies a line read by *fgets* in which *NODES* are the first five characters. The *if* statement that follows the *while* loop detects which of these cases is relevant. Note how the associated *else* displays a suitable message and sets *return_code* to 2 if the end of the file is detected.

Assuming that the start of the *NODES* table has been found, *fgets* is used to read the next (blank) line, prior to the program reading the numerical values from the *NODES* table. The actual reading process starts by initializing the nodes counter, i, to zero. *get_nodes* then uses another *while* loop which increments *i* each time a line has been successfully read. The reading process involves two steps. The first uses *fgets* to read a whole line, again storing it in *line*. This allows end of file to be detected. Next, *sscanf* is used to read values out of *line*, using the integer returned by *sscanf* to check that the required three values have been read. Note that the complete process for reading a node definition is contained in the *while* loop continuation statement; and that the loop controls just a single statement, that increments the nodes counter, *i*.

The *while* loop terminates either if *fgets* returns a *NULL* pointer value, or if *sscanf* returns a value other than 3. If *fgets* returns a *NULL* pointer, this means that the end of the file has been found. This is communicated back to *main* by setting *return_code* to 3. Conversely, if *sscanf* returns a value other than 3, the end of the *NODES* table has been reached and *get_nodes* has successfully completed its task. This is communicated back to *main* by *return_code* still having its initialized value of zero. In either case, the number of nodes that have been read is stored in the *no_nodes* member of *mesh*. On return to *main*, the value returned by *get_nodes* is stored in a variable, also called *return_code*, that is local to *main*. This is used to display an error message if *get_nodes* was unsuccessful, or to call the *display_nodes* function. If *display_nodes* is called, *main* passes *mesh* to it by value, with *display_nodes* making a copy, also called *mesh*. The *display_nodes* function then uses a *for* loop to display the nodes data on the screen.

Tutorial 5.7
Implement Program 5.3 and make notes on its operation.

Tutorial 5.8
Modify Program 5.3 to read and display the numerical data in both tables contained in the input file.

Chapter review

The new material introduced in this chapter is relatively small. However, it builds on previously discussed features of C to provide a wide range of flexible and robust methods for working with files. Streams are the 'core technology' provided by C for connecting a program to one or more files. Unlike the standard streams, *stdin* and *stdout*, any stream needed for working with a file must be declared and opened using appropriate statements in the program. Streams are actually pointers to data structures of type *FILE*. The *fopen* function is used to open a stream between the program and a named file, the stream being the pointer returned by *fopen*. When opening a stream, *fopen* also needs to know the type of connection – read '*r*', write '*w*' or append '*a*'. When no longer required, streams should be closed (breaking the link between the program and the file) using the *fclose* function. Some of the more generally useful features of formatting codes and formatting code modifiers have also been discussed. Specifically, methods for controlling the precision of outputted data and use of the * modifier to ignore chosen inputs. It should be noted, however, that C provides other modifiers that have not been discussed here and, to investigate these, the reader should consult HELP in their programming environment. Functions similar to *fprintf* and *fscanf* have been introduced that allow character strings to be assembled using *sprintf*, and allow data to be read from character strings using *sscanf*. In using both of these functions, the first argument passed to them is the name of the relevant character string, rather than the name of a stream. The *sscanf* function is frequently used in conjunction with the *fgets* function, the latter allowing whole lines, rather than particular data items to be read from files. It has also been shown that *fgets*, in combination with *sscanf*, string handling functions and *while* loops, can provide flexible navigation within input files. In turn, this enables potential errors to be anticipated and accommodates situations where the sequence and/or volume of input data are not known before reading commences.

6

Dynamic Memory Management and Linked Lists

6.1 Introduction

C provides a collection of functions that allow variables to be created and destroyed whilst a program is running. What this means is that sections of memory can be reserved or allocated and used to store data when required. When the data stored in these locations is no longer required, the allocated memory can be released or freed, becoming available for possible re-use at some other time. This method of memory management is called '**dynamic**' because the C program decides when to use it. In contrast, the use of arrays is called '**static**' memory management because array sizes are fixed before the program runs.

As seen in previous chapters, using arrays in programs that may need to process varying amounts of data always carries the risk that the arrays are not big enough to hold all of the data. In large programs this is a serious problem that can be overcome through the use of dynamic memory management. In using dynamic memory management, it is typical to design methods of storing data by dynamically creating data structures containing several member variables. Some of these members are used to store the processed data and other members are pointers that can store the addresses of other data structures. Using these pointers, as many data structures as required can be chained together forming a linked list.

This chapter presents the essential facilities for dynamic memory management, demonstrating them through various examples. Attention then moves on to **linked lists** and their practical use.

6.2 Essential facilities for dynamic memory management

To use any dynamic memory management facilities, a C program must include a standard library to provide the necessary function prototypes. This library is either *stdlib.h* or *alloc.h*, depending on the programming environment being used.

C provides the following functions:

```
void *malloc(no_bytes)
void *calloc(no_blocks, no_bytes)
void *realloc(current_storage_ptr, no_bytes)
void free(current_storage_ptr)
```

where:

- *void* * means the function returns a pointer (an address in memory), but the pointer does not have a data type;
- *no_bytes* is an integer that specifies the number of bytes to be allocated as a single block of memory;
- *no_blocks* is an integer that specifies the number of blocks of memory to be allocated;
- *current_storage_ptr* is a pointer to a block of memory that is currently allocated.

The *malloc* function allocates a single block of memory and the *calloc* function allocates a number of contiguous blocks. *malloc* returns a pointer of type *void* that holds the address of the first byte in the allocated block. *calloc* returns a pointer of type *void* that holds the address of the first byte in the first allocated block. The *realloc* function changes an amount of memory that has already been allocated in a block. Thus, a pointer to the first byte of an allocated block is passed to *realloc*, along with the new number of bytes to be allocated. The *free* function removes or de-allocates a block of memory that has been previously allocated using either *malloc* or *calloc*.

The contents of a dynamically allocated block of memory can only be accessed by reference, using the pointer that is returned from *malloc* or *calloc*. If *no_bytes* is an explicit integer value, *malloc* or *calloc* will structure the allocated blocks so that individual bytes can be accessed. However, it is more usual to specify the size of a required block in terms of a particular data type using the *sizeof* operator (Section 2.5) and to then convert the returned pointer to a pointer of the same data type using the cast operator (Section 2.5). When this is done, the memory within a block can then be accessed in terms of the specified data type. For example:

```
int *integer_ptr;
integer_ptr = (int *)malloc(sizeof(int));
*integer_ptr = 5;
```

Above, *malloc* allocates enough memory (2 bytes) to store an *int*, returning its address in a pointer of type *void*. The returned pointer is then cast to be a pointer of type *int* and the address that it holds is assigned to *integer_ptr*. Finally, the value 5 is stored in the allocated memory using the 'contents of' operator (Section 2.5). Also consider,

```
double *double_ptr;
double_ptr = (double *)malloc(sizeof(double));
*double_ptr = 8.4;
```

Here, *malloc* allocates enough memory (8 bytes) to store a variable of type *double*, returning its address in a pointer of type *void*. The returned pointer is cast to be a pointer of type *double* and the address that it holds is assigned to *double_ptr*. After this, the 'contents of' operator is used to store the value 8.4 in the allocated memory.

A more typical example is shown below, where sufficient memory is allocated to store a data structure.

```
struct triangle
  {
  double x[3];
  double y[3];
  double area;
  };
struct triangle *triangle_ptr;
triangle_ptr = (struct triangle *)malloc(sizeof(struct triangle));
```

The first six lines, above, specify the template for a structure called *struct triangle*. In line 7 the template is used as a data type to declare a pointer called *triangle_ptr*. In the final line, *malloc* allocates enough memory (56 bytes) to store a variable of type *struct triangle*, returning its address in a pointer of type *void*. The returned pointer is then cast to be a pointer of type *struct triangle* and the address that it holds is copied to *triangle_ptr*.

In these three examples, it is important to note that the data type of the pointer obtained after using the cast operator and the data type passed to the *sizeof* operator are the same.

Tutorial 6.1

Convert the following into working programs, correcting the single mistake contained in each and displaying the result on the screen.

a)   *int *integer_ptr;*
     integer_ptr = (float *)malloc(sizeof(int));
     *integer_ptr = 5;

b)   *double *double_ptr;*
     double_ptr = (double *)malloc;
     *double_ptr = 8.4;

Tutorial 6.2

Write a program that reads, stores and displays the following data: 125.7, 95 and *'disc'*. The data must be stored in a single data structure using member variables of the appropriate type. Use dynamic memory management to create the data structure.

6.3 Simple applications of dynamic memory management

Program 6.1 shows a simple example of how memory management functions can be used in practice. The objective of this program is to calculate the area of a rectangle, which is defined by the *x,y* co-ordinates of its lower left and upper right corners. The co-ordinates of a corner are stored in a data structure, called *corner*. The *malloc* function is used to allocate storage for two variables of type *corner*, the addresses of these variables being stored in *lower_left_ptr* and *upper_right_ptr*.

/ Program 6.1 - Calculating the area of a rectangle */*

```
#include <stdio.h>
#include <stdlib.h>

int main(void)
{
struct corner
  {
  double x;
  double y;
  };
struct corner *lower_left_ptr, *upper_right_ptr;
double area;

lower_left_ptr = (struct corner *)malloc(sizeof(struct corner));
upper_right_ptr = (struct corner *)malloc(sizeof(struct corner));

lower_left_ptr->x = 0.0;
lower_left_ptr->y = 0.0;
upper_right_ptr->x = 10.0;
upper_right_ptr->y = 10.0;
area = (upper_right_ptr->x - lower_left_ptr->x) *
        (upper_right_ptr->y - lower_left_ptr->y);

fprintf(stdout," area = %lf\n", area);
return(0);
}
```

Program 6.1 uses two calls to the *malloc* function, the first to allocate the structure for the lower left corner and the second to allocate the structure for the upper right corner. Note in these statements that *struct corner* is passed to *sizeof* to obtain the number of bytes needed to hold one instance of *struct corner*. The number of bytes returned by *sizeof* is then passed as an argument to *malloc*, which allocates a memory block of the correct size and then returns a pointer to the first byte in the block. Finally, the cast operator converts the data type of the returned pointer to *struct corner* before its value is copied to *lower_left_ptr* in the first call to *malloc* and to *upper_right_ptr* in the second call.

Having allocated the memory required for each structure, their member variables are then assigned co-ordinate values that define the rectangle. Note that members of each structure are accessed by using the relevant pointer. This is the only way to access members of dynamically allocated structures. It is not possible to fully qualify members of allocated data structures because such structures do not have names. This is demonstrated again in the statement used to calculate the area of the rectangle.

To further develop the use of dynamic memory allocation, consider Program 6.2, which is a modified version of Program 5.2. Program 6.2 aims to demonstrate how blocks of memory can be dynamically allocated within functions and how those functions can return pointers to allocated memory back to the function that called them. This example also demonstrates that dynamically allocated memory can be passed by reference to functions, in just the same way as explicitly declared variables. A detailed list of the changes that have been made to obtain Program 6.2 from Program 5.2 is given after the program statements.

In Program 6.2 templates for the *files* and *triangle* data types are declared external to each function, so that all of the functions share a common understanding of these data types. *main* declares two pointers, *io_ptr* and *example_ptr*, using these data types. It is intended that the *read_filenames* function will allocate memory needed to store the filenames and that the *read_points* function will allocate memory needed to store the points that define a triangle. Both functions will return the addresses of this allocated memory back to *main* which will store them in the previously declared pointers. The prototype statements in *main* for the *read_filenames* and *read_points* functions are consistent with this.

Looking at the argument lists in the prototype statements for *read_filenames* and *read_points*, no arguments are passed to *read_filenames*, so its argument list contains *void*. The single argument passed to *read_points* is a character string intended to contain the name of the file where the data for each point are stored. Similarly, in the prototype statement for *write_area*, the second variable is a character string that will contain the name of the output file. The first two executable statements in *main* call the *read_filenames* and *read_points* functions, assigning their returned pointer values to *io_ptr* and *example_ptr*, respectively.

Now look at the *read_filenames* function, which declares its own local pointer, called *io_ptr*, to store its copy of *io_ptr* passed to it from *main*. The *malloc* function is used to allocate a block of memory of the correct size to hold an instance of the *struct files* data structure. The pointer returned by *malloc*, containing the address of this block of memory, is first cast to the *struct files* data type and then its value is copied into *io_ptr*. After reading the names of the files from the user (note how the file name members are accessed using the 'address of' operator with the pointer to the allocated data structure) the value of *io_ptr* is returned to *main*.

Looking at the *read_points* function, it can be seen that the character string passed to it is copied into the character string, *input_filename*, appearing in its argument list. Also, there are two other declaration statements, the first creating a stream called *input* and the second for a pointer of data type *struct triangle*. The *malloc* function is used to allocate a block of memory in which to store an instance of the *struct triangle* data type. The address of the block allocated by *malloc* is stored in *triangle_ptr*. Following this, *fopen* is used to connect the program to the input file using the *input* stream. Note, in the subsequent calls to *fscanf*, that members of the *triangle* data structure are identified using *triangle_ptr*, rather than a structure name. After the last call to *fscanf*, the *fclose* function breaks the link between the program and the input file and the function returns the value of *triangle_ptr* to *main*. The remainder of the program is the same as Program 5.2.

```
/* Program 6.2 - Calculating the area of a triangle                      */
/*                                                                        */
/* Main function for program 6.4.                                        */
/* Calculate the area of a triangle which is defined by three pairs of x,y */
/* co-ordinates, supplied by the user.                                   */
/* Demonstrates the dynamic allocation of memory within functions, the */
/* return of pointers to allocated memory and the passing of allocated   */
/* memory to functions by reference.                                     */


#include <stdio.h>
#include <math.h>
#include <stdlib.h>
```

```
struct triangle
  {
  double x[3];
  double y[3];
  double area;
  };

struct files
  {
  char input_filename[101],
      output_filename[101];
  };

int main(void)
{
struct files *io_ptr;
struct triangle *example_ptr;

struct files *read_filenames(void);
struct triangle *read_points(char[]);
int calculate_area(struct triangle *);
int write_area(struct triangle *, char[]);

io_ptr = read_filenames();
example_ptr = read_points(io_ptr->input_filename);
calculate_area(example_ptr);
write_area(example_ptr, io_ptr->output_filename);
return(0);
}
```

```
/* Function: read_filenames; Used in program 6.2.              */
/* Reads two file names as char strings into an allocated structure.  */
/* The function returns a pointer to the allocated structure.    */

struct files *read_filenames(void)
{
struct files *io_ptr;
```

```
io_ptr = (struct files *)malloc(sizeof(struct files));
fprintf(stdout," Input file name:");
fscanf(stdin," %s",io_ptr->input_filename);

fprintf(stdout," Output file name:");
fscanf(stdin," %s",io_ptr->output_filename);
return(io_ptr);
}
```

```
/* Function: read_points; Used in program 6.2.                    */
/* Reads x,y co-ordinates of triangle vertices into an allocated structure. */
/* A pointer to the allocated structure is returned to the calling function.  */

struct triangle *read_points(char input_filename[])
{
FILE *input_file;
struct triangle *triangle_ptr;

triangle_ptr = (struct triangle *)malloc(sizeof(struct triangle));
input_file = fopen(input_filename, "r");

fscanf(input_file," %lf %lf", &triangle_ptr->x[0], &triangle_ptr->y[0]);
fscanf(input_file," %lf %lf", &triangle_ptr->x[1], &triangle_ptr->y[1]);
fscanf(input_file," %lf %lf", &triangle_ptr->x[2], &triangle_ptr->y[2]);
fclose(input_file);
return(triangle_ptr);
}
```

```
/* Function: calculate_area; Used in program 6.2.     */
/* Calculates area of triangle defined by (x,y)       */
/* co-ordinates supplied in structure pointed to by    */
/* triangle_ptr.                                        */
```

```
int calculate_area(struct triangle *triangle_ptr)
{
double  a,       /* distance between points 1 and 2      */
        b,       /* distance between points 2 and 3      */
        c,       /* distance between points 3 and 1      */
        s;       /* perimeter / 2                        */

a = sqrt((triangle_ptr->x[1] - triangle_ptr->x[0]) *
         (triangle_ptr->x[1] - triangle_ptr->x[0]) +
         (triangle_ptr->y[1] - triangle_ptr->y[0]) *
         (triangle_ptr->y[1] - triangle_ptr->y[0]));

b = sqrt((triangle_ptr->x[2] - triangle_ptr->x[1]) *
         (triangle_ptr->x[2] - triangle_ptr->x[1]) +
         (triangle_ptr->y[2] - triangle_ptr->y[1]) *
         (triangle_ptr->y[2] - triangle_ptr->y[1]));

c = sqrt((triangle_ptr->x[0] - triangle_ptr->x[2]) *
         (triangle_ptr->x[0] - triangle_ptr->x[2]) +
         (triangle_ptr->y[0] - triangle_ptr->y[2]) *
         (triangle_ptr->y[0] - triangle_ptr->y[2]));

s = (a + b + c)/2.0;
triangle_ptr->area = sqrt(s*(s-a)*(s-b)*(s-c));
return(0);
}



/* Function: write_area; Used in program 6.2.              */
/* Writes calculated area of a triangle to a file.         */
/* Name of output file is supplied by calling function.    */

int write_area(struct triangle *triangle_ptr, char output_filename[])
{
FILE *output_file;

output_file = fopen(output_filename,"w");
fprintf(output_file,"Area of triangle = %f\n", triangle_ptr->area);
fclose(output_file);
return(0);
}
```

The following changes have been made to develop Program 6.2 from Program 5.2.

Before main

Dynamic memory allocation functions are made available to the program by using the *#include <stdlib.h>* statement.

In main

Variables of the *struct files* and *struct triangle* data types are no longer declared, but the declarations for pointers to variables of these types are retained.

The prototype statements for the *read_files* and *read_points* functions are both changed to specify that each function returns a pointer of the relevant data type.

The prototype statement for *read_files* also specifies that no arguments are passed to the function. The implication of this and the previous change to the prototype statement for this function is that *read_files* will use *malloc* to create an instance of the *struct files* data type and will return a pointer to it, rather than sharing access to an empty copy passed to it.

In the prototype statement for *read_points* the pointer to the *triangle* data structure has been removed. Again, the implication of this change to the *read_points* prototype statement is that the function will use *malloc* to create an instance of the *struct triangle* data type and will return a pointer to it.

Consistent with *read_files* and *read_points* returning pointers to the relevant data structures, the explicit assignments of structure addresses to *example_ptr* and *io_ptr* in Program 5.2 have been removed.

In read_filenames

The function name statement now indicates that the function returns a pointer of data type *struct files* and that no arguments are passed to the function.

A declaration statement has been introduced for the pointer, *io_ptr*.

A further statement is introduced that calls *malloc* to create a structure of the *struct files* data type, storing its address in *io_ptr*.

The *return* statement now returns *io_ptr*, rather than zero.

In read_points

The function name statement now indicates that the function returns a pointer of the *struct triangle* data type and that only the name of the input file is passed to the function.

A declaration statement has been introduced for the pointer, *triangle_ptr*.

A further statement is introduced that calls *malloc* to create a structure of the *struct triangle* data type, storing its address in *triangle_ptr*.

The *return* statement now returns *triangle_ptr*, rather than zero.

Tutorial 6.3
Implement Program 6.1 and make brief notes on its operation.

Tutorial 6.4
Implement Program 6.2 and make brief notes on its operation.

6.4 Linked lists

A linked list is a collection of data structures that are connected together using pointers. For this to be possible, each structure in the list must contain a member variable that is a pointer. By storing the address of one data structure in a pointer that is a member of another data structure a link is made between the two structures. The most simple data structure that can usefully form part of a linked list is one that contains a single basic variable and a pointer variable, as shown below.

```
struct integer_value
    {
    int value;
    struct integer_value *next;
    };
struct integer_value *first_ptr, *second_ptr;
```

```
first_ptr = (struct integer_value *)malloc(sizeof(struct integer_value));
second_ptr = (struct integer_value *)malloc(sizeof(struct integer_value));

first_ptr->value = 10;
first_ptr->next = second_ptr;

second_ptr->value = 20;
second_ptr->next = NULL;
```

The linked list produced by the example code, above, is shown graphically in Figure 6.1.

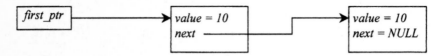

Figure 6.1 *A linked list containing two data structures*

In this example the first task is to define a template for a data structure called *struct integer_value*, along with two pointers, *first_ptr* and *second_ptr*, of the same data type. Note that the template contains two members, the first is an integer variable called *value* and the second is a pointer, called *next*, whose data type is the same as the structure in which it is a member. The code then calls *malloc* twice to create two structures of type *integer_value*. The addresses of the structures, returned from the calls to *malloc* are stored in *first_ptr* and *second_ptr*. After assigning a value of 10 to member *value* in the first structure, the address of the second structure, stored in *second_ptr*, is copied to the member, *next*, in the first structure. This makes a link between the first and second data structures. The pointer member, *next*, of the second structure is assigned the *NULL* value, which is used to indicate the second structure is the last in the linked list.

Program 6.3 demonstrates how linked lists are typically used. This program is a modified version of Program 4.6, which prompts the user to name a range of fruit. The latter program was limited in that a maximum of 20 names could be held in an array of character strings. If the user supplied more than 20 names, the behaviour of the program would be unpredictable because names after the 20th would overwrite memory used to store other variables. There are two possible solutions to this problem. The first involves the use of an array that is big enough to hold a large number of names. This is not a good solution because how big is 'big enough'? At best, this results in a waste of memory and, at worst, the original problem re-occurs. A much better solution is to store the names in a linked list, where each new name is stored in a new data structure. In addition to the statements in Program 6.3, Figure 6.2 (page 130) provides an overview of the steps taken within the program to build the linked list.

```c
/* Program 6.3 - The fruit program yet again */

#include <stdio.h>
#include <string.h>
#include <stdlib.h>

int main(void)
{
char reply[20];
struct fruit
  {
  char name[20];
  struct fruit *next;
  };
struct fruit *first_ptr = NULL, *new_structure_ptr, *current_fruit_ptr;

do
  {
  fprintf(stdout,"Name of fruit or 'end' ?:");
  fscanf(stdin,"%s", reply);
  if (strncmp(reply,"end",3) != 0)
    {
    new_structure_ptr = (struct fruit *)malloc(sizeof(struct fruit));
    if (first_ptr != NULL)
        current_fruit_ptr->next = new_structure_ptr;
    else
        first_ptr = new_structure_ptr;
    current_fruit_ptr = new_structure_ptr;
    strcpy(current_fruit_ptr->name, reply);
    current_fruit_ptr->next = NULL;
    }
  }
while(strncmp(reply,"end",3) != 0);
```

```
fprintf(stdout," You have named the following fruit:\n");
current_fruit_ptr = first_ptr;
while(current_fruit_ptr != NULL)
   {
   fprintf(stdout," %s\n", current_fruit_ptr->name);
   current_fruit_ptr = current_fruit_ptr->next;
   }
return(0);
}
```

In Program 6.3 a template data structure is defined to hold the name of a single fruit. This structure also contains a pointer that can store the address of another data structure of the same type. Three other pointers of the same data type are also declared (Figure 6.2(a)):

- *first_ptr* is used to store the address of the first structure in the linked list. It is initialized to a *NULL* address to indicate that the linked list is initially empty.
- *new_structure_ptr* is used to temporarily store the address returned by *malloc*.
- *current_fruit_ptr* is used to access the latest structure that has been added to the list.

As in Program 4.6, all of the statements used in Program 6.3 to obtain data from the user and store it in the linked list is carried out inside a *do-while* loop, whose *continuation expression* tests for the user entering '*end*'. Having prompted for and read user input, an *if* statement is used to decide whether the user has supplied a fruit name. If they have, a *fruit* data structure is allocated using *malloc* and its address is temporarily stored in *new_structure_ptr*. Following this, a test is carried out to decide if this is the first structure that has been allocated. Let's assume that it is, so the *else* part of the test is executed and the structure address is stored in *first_ptr*. Next, the structure address is also stored in *current_fruit_ptr*, so that the new structure becomes the one currently being processed. At this point, the user input is copied into the *name* member of the structure and the pointer member, *next*, is set to *NULL*. This latter statement ensures that if no more fruit names are supplied then the current structure becomes the last structure in the list.

Figure 6.2(b) shows that one data structure has now been allocated and that each of the previously declared pointer variables contain its address. The *continuation expression* in the *while* statement is now tested, which takes the program back up to the *do* at the top of the loop because the user did not enter *'end'*. The user is again prompted for input. Assuming that they supply a second fruit name, another data structure is allocated. Since *first_ptr* now holds the address of the first allocated data structure, the test expression in the *if* statement is now TRUE causing the first part of the *if-else* statement to be executed. This uses the present value of *current_fruit_ptr*, now dotted in Figure 6.2(c), to store the address of the new data structure in the member *next* of the previous structure.

Following this, *current_fruit_ptr* is updated to make the newly allocated structure the current one, prior to storing the user supplied name and setting *next* in the current structure to *NULL*. Figure 6.2(c) shows the state of the linked list and its associated pointers at the end of the second pass through the *do-while* loop. Similarly, Figure 6.2(d) shows the situation after the user has supplied a third name. Note, again, that *current_fruit_ptr* is used to change the value of *next* in the previously allocated structure, prior to being assigned the address of the new structure. Finally, supposing that the user entered five names, followed by *'end'*, the linked list and its pointers will be as shown in Figure 6.2(e) and the program will leave the *do-while* loop.

In order to display the full list of fruits, the address stored in *first_ptr* is now copied to *current_fruit_ptr*, so that *current_fruit_ptr* points to the start of the list. A *while* loop is then used to, firstly, print the name stored within the current structure and, secondly, update *current_fruit_ptr* by assigning to it the value of *next* in the current structure. The *while* loop terminates when *current_fruit_ptr* has been assigned a *NULL* value, indicating that the end of the list has been reached.

In comparison to Program 4.6, Program 6.3 has improved functionality through its increased robustness, brought about by using dynamic, rather than static, memory management. Comparing the statements in each program, this is achieved through a small increase in the complexity of the code, which is based on a significant difference in thinking about memory use and software reliability.

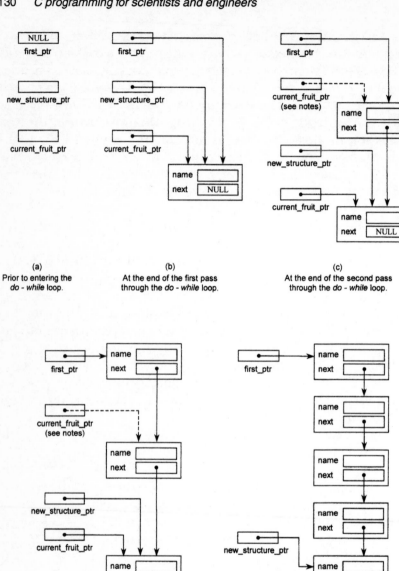

(a)
Prior to entering the
do - while loop.

(b)
At the end of the first pass
through the *do - while* loop.

(c)
At the end of the second pass
through the *do - while* loop.

(d)
At the end of the third pass
through the *do - while* loop.

(e)
After the *do - while* loop when a total
of five names have been supplied.

Figure 6.2 *The development of a linked list to store fruit names in Program 6.3*

Tutorial 6.5
Implement Program 6.3 and make notes on its operation.

Tutorial 6.6
Modify Program 6.3 to read any number of fruit names from a user specified file and write the list of names to another user specified file.

Chapter review

Dynamic memory management is a very powerful technique for producing robust software that is efficient in its use of system resources. This chapter has introduced the essential elements of dynamic memory management through the use of small example programs. It could be argued that the use of dynamic memory management in such small programs may not be appropriate. However, this chapter is also intended as a bridge between such programs and their much larger counterparts that are built to solve complex technical problems. Such software, possibly consisting of hundreds of functions and hundreds of thousands of statements typically depends on the use of dynamic memory management introduced here.

Dynamic memory management is generally perceived as a fairly advanced technique, mainly for the following reasons. Firstly, its use relies on a good understanding of other C facilities, such as data structures, pointers, functions, loops and decision making. Secondly, to design software that will use dynamic memory management, it is necessary to develop a good mental picture of relationships between individual items of data and between the data and the instructions needed to process it.

Appendix: Typical Examination Questions

Chapter 1

1. Implement the following program that will display the default size (in bytes) of each basic C data type on your system. Invent a bar chart using the results to show the relative amounts of memory needed for each data type. Note that *sizeof(...)* is an operator provided by C that works out how many bytes are needed to store a variable of any data type. Imagine, below, that in *sizeof(short int)* etc., the number of bytes worked out by *sizeof* replaces *sizeof(short int)* etc. in the *fprintf* argument list.

```
#include <stdio.h>
int main(void)
{
fprintf(stdout," short int needs: %d bytes\n", sizeof(short int));
fprintf(stdout,"      int needs: %d bytes\n", sizeof(int));
fprintf(stdout," long int needs: %d bytes\n", sizeof(long int));
fprintf(stdout,"    float needs: %d bytes\n", sizeof(float));
fprintf(stdout,"   double needs: %d bytes\n", sizeof(double));
fprintf(stdout,"     char needs: %d bytes\n", sizeof(char));
fprintf(stdout,"int pointer needs: %d bytes\n", sizeof(int *));
fprintf(stdout,"double pointer needs: %d bytes\n", sizeof(double *));
return(0);
}
```

2. Why are the following statements false?
   a) A character string, declared as *char string[7]*, is long enough to store the word 'halibut'.
   b) *char word[7]* = *'reggae';* is a correct initialization of the character string, *word*.
   c) The value 1.768 can be stored in a variable of type *int*.
   d) A variable of type *short int* can be used to store the value 48927.
   e) A variable of type *unsigned int* can be used to store the value −1.
3. Implement a program that uses variables of the most appropriate data type to store the values shown below. The program must initialize the variables with the given values and then use one or more calls to *fprintf* to display the values on the screen. The values are:

   21.6, −32769, 120, −120, 9.8475432877e+9, chair,
   chair_number_26, A

4. Use the HELP facility in your programming environment to investigate the following:

   stdio.h, fprintf, fscanf

5. Implement a program that reads data from the keyboard into the following data structure and displays it on the screen:

   ```
   struct component
     {
     int identity;
     int order_no;
     double weight;
     char colour[11];
     };
   ```

6. The following program contains five errors. Fix the errors and demonstrate that the program works correctly.

   ```
   include <stdio.h>
   int main(void)
   {
   int x = 19.7;
   char a[1];
   ```

fprintf(stdout, A variable, x, has been declared\n");
fprintf(stdin, Enter a character string (max. 9 symbols):");
fscanf(stdout, %s", a);
fprintf(stdout, The string and the variable are %s and %d, respectively\n",
a, x);
return(0);
}

Chapter 2

7. Linear steady-state heat conduction in a bar is described by:

$$q = -kA \frac{(T_2 - T_1)}{(x_2 - x_1)}$$

where

q = Heat flux (W/m^2)
k = Thermal conductivity (W/mK)
A = Cross sectional area (m^2)
T_1, T_2 = Temperatures at each end of the bar (deg.K)
x_1, x_2 = Locations of the ends of the bar (m)

Write a program that implements the given equation. The program should prompt for and read values of k, A, x_1, T_1, x_2, T_2 from the user, indicating the required units. The calculated value of q should be displayed on the screen as part of an appropriate message to the user. Choosing values for the inputs, calculate q by hand and demonstrate that your program works correctly.

8. Gross margin, net profit and their percentage values are important measures of financial performance, and can be defined as:

$$\text{gross\_margin} = \text{sales} - \text{variable\_costs} \text{ (\pounds)}$$

$$\text{percentage\_gross\_margin} = \frac{\text{gross\_margin}}{\text{sales} * 100}$$

$$\text{net\_profit} = \text{gross\_margin} - \text{overhead\_costs} \text{ (\pounds)}$$

$$\text{percentage\_net\_profit} = \frac{\text{net\_profit}}{\text{sales} * 100}$$

Write a program that implements the given equations. The program should prompt for and read values of *sales*, *variable_costs* and *overhead_costs* from the user, indicating the required units. The calculated values should be displayed on the screen as parts of appropriate messages to the user. Choosing values for the inputs, calculate the financial measures by hand and demonstrate that your program works correctly.

9. According to the simple bending equation, when a beam of rectangular cross section is bent the maximum and minimum tensile stresses on its lower and upper surfaces, respectively, are given by:

$$\sigma = \pm \frac{Md}{2I}$$

where

σ = Maximum and minimum tensile stress (N/m²)
d = Depth of beam (m)
M = Applied bending moment (Nm)
I = Second moment of area (m⁴)

For a beam having a rectangular cross section, I is given by

$$I = \frac{bd^3}{12} \text{ (m}^4\text{)}$$

where
b = Breadth of beam (m)

Write a program that implements the given equations. The program should prompt for and read values of M, b and d from the user, indicating the required units. The calculated maximum and minimum stresses should be displayed on the screen as parts of appropriate messages to the user. Choosing values for the inputs, calculate the maximum and minimum stresses by hand and demonstrate that your program works correctly.

10. The area of a triangle enclosed by three points $P_1(x_1, y_1)$, $P_2(x_2, y_2)$ and $P_3(x_3, y_3)$ is given by:

$$\text{Area} = \sqrt{s(s-a)(s-b)(s-c)}\ (\text{m}^2)$$

where

a = Straight line distance between P_1 and P_2 (m)
b = Straight line distance between P_2 and P_3 (m)
c = Straight line distance between P_3 and P_1 (m)

and

$$s = \frac{a+b+c}{2}\ (\text{m})$$

Write a program that implements the given equations. The program should prompt for and read the x and y values of three points from the user, indicating the required units. The calculated area should be displayed on the screen as part of an appropriate message to the user. Choosing values for the inputs, calculate the area by hand and demonstrate that your program works correctly. Hint: use Pythagoras's Theorem to calculate a, b and c.

Chapter 3

11. Write and implement a program that reads two integers from the user, adds them together and displays the answer on the screen. The program must consist of functions for each major activity (reading, processing and writing).
12. Re-implement the program in Chapter 2, Question 7 so that it uses separate functions for the reading, calculating and writing tasks.
13. Re-implement the program in Chapter 2, Question 8 so that it uses separate functions for the reading, calculating and writing tasks.
14. Re-implement the program in Chapter 2, Question 9 so that it uses separate functions for the reading, calculating and writing tasks.

Chapter 4

15. Write a program that displays the alternative text strings shown at the end of the question and prompts the user to enter any one of them. The program should use nested *if-else* statements to compare the user's input against the list of valid strings and print a message indicating which string the user has supplied. The program must also print an error message if the user enters a string that does not appear in the list. The valid strings are:

    'ABC' 'DEF' 'GHI' 'JKL' 'MNO' 'PQR' 'STU' 'VWX' 'YZ'

16. Repeat Question 15, replacing the *if-else* construct by a *switch* construct.

17. Re-write Program 2.2 from Chapter 2 so that the text 'TRUE' and 'FALSE' appear in the output, rather than the numerical values 1 and 0, respectively.

18. Write a program that can repeatedly prompt the user to enter the name of any of the data types considered in Sections 1.2, 1.3 and 1.4, e.g. *int*, *float*, etc. For any valid data type supplied by the user, the program should display the following:

    - The name of the data type.
    - The number of bytes that it uses.
    - The upper and lower limits of values that can be stored in variables of that type.
    - Where appropriate, the precision of the values that can be stored.

    Your program should store the relevant values, found from the notes in Chapter 1, in suitable arrays rather than 'hard coding' them into calls to *fprintf*. The program should display a suitable error message if the user enters invalid input, and should allow the user to stop the program by entering a suitable command, such as '*end*'.

19. Write a program that initializes an array with several alphabetical characters. The program should then use a *while* loop to display characters randomly selected from the array and, for each character, prompt the user to enter a word beginning with that character. The program should check that the user has supplied a correct input and display an appropriate message if their input is not correct. The user should be able

to end the program by entering '*end*'. Hint: it may be better to work out how to change Program 4.5 than to write a completely new program. Another hint: think carefully about the user being able to enter '*end*' as a valid word beginning with '*e*'.

20. Modify the program for either Tutorial 4.7 or 4.8 so that the *switch* construct operates inside a *while* loop. This should allow the user to select options 0, 1 and 2 in any sequence. Introduce a third option that enables the user to stop the program.

Chapter 5

21. Modify the program in Chapter 3, Question 14 (derived from Chapter 2, Question 9) so that it can read M, b and a range of d values from a file. The function used to perform the calculation should now use a loop to carry out the calculation for each d value. The program should also write a table containing the d values and associated maximum and minimum stresses, accurate to three decimal places, to an output file. The program should read the names of the input and output files from the user.

22. Modify the program written for Chapter 4, Question 19 so that words beginning with different characters are written to different files. If the user is prompted to supply more than one word beginning with the same letter, each word after the first should be appended to the relevant output file. File names should be constructed by the program when they are needed.

Chapter 6

23. Modify the program in Tutorial 6.6 so that the list of fruit names is written to a file in reverse order. Hint: use two pointers in each data structure, one 'pointing' forwards and the other 'pointing' backwards.

24. Using the relevant parts of Programs 6.1 and 6.3 as a starting point, write a program that uses a linked list to store the vertices of a two-dimensional polygon having any number of sides. The vertex data is to be supplied via the keyboard and the user must not need to specify how many sides make up the

polygon before they input the vertex data. Having read all of the vertex data, the program should write it to a file.

25. Making use of the relevant parts of the program in Question 24, write a program that reads the polygon vertex data from the file into a linked list. The program should count the number of vertices and then prompt the user to select one of them. Starting with the user-selected vertex, the program should write all of the vertex data to a second file. Hint: you will need to think about how to 'wrap' the end of the linked list around to the beginning.

26. Compare Programs 4.6 and 6.3, concentrating on the limitations of the former, discussed in Section 4.8, and the ways in which they are overcome in Program 6.3.

Background and Rationale of the Series

This new series has been produced to meet the new and changing needs of students and staff in the Higher Education sector caused by firstly, the introduction of 15 week semester modules and, secondly, the need for students to pay fees.

With the introduction of semesters, the 'focus' has shifted to module examinations rather than end of year examinations. Typically, within each semester a student takes six modules. Each module is self-contained and is examined/assessed such that on completion a student is awarded 10 credits. This results in 60 credits per semester, 120 credits per year (or level to use the new parlance) and 360 credits per honours degree. Each module is timetabled for three hours per week. Each semester module consists of 12 teaching weeks, one revision week and two examination weeks. Thus, students concentrate on the 12 weeks and adopt a compartmentalized approach to studying.

Students are now registered on modules and have to pay for their degree per module. Most now work to make ends meet and many end up with a degree and debts. They are 'poor' and unwilling to pay £50 for a module textbook when only a third or half of it is relevant.

These two things mean that the average student is no longer willing or able to buy traditional academic text books which are often written more for the ego of the writer than the needs of students. This series of books addresses these issues. Each book in the series is short, affordable and directly related to a 12 week teaching module. So modular material will be presented in an accessible and relevant manner. Typical examination questions will also be included, which will assist staff and students.

However, there is another objective to this book series. Because the material presented in each book represents the state-of-the-art practice, it will also be of interest to professional engineers in industry and specialist practitioners. So the books can be used by engineers as a first source reference that can lead onto more detailed publications.

Therefore, each book is not only the equivalent of a set of lecture notes but is also a resource that can sit on a shelf to be referred to in the distant future.

Index